THE BEST

FIRE PICTURES

OF 1984

THE BEST
FIRE PICTURES
OF 1984

Addison C. Getchell and Son, Inc.

Printed in the United States of America.

First Edition.

Editor: Henry M. Quinlan

Design: Designworks, Inc.
5 Bridge Street
Watertown, Massachusetts

Table of Contents

Introduction

This book is an attempt to bring together in one place the best photographs of fires and the people who fight fires for the year 1984. The book contains a diverse selection of pictures from all sections of the country and together they produce a collective portrait of firefighting in America. Each photographer presents his or her view of a particular fire. Some emphasize the firefighters; some the fire equipment; some the victims, and some the fires themselves.

Fires have fascinated man for all of recorded history. Man's perception of fires have progressed from early drawings through vivid paintings to today's spectacular photography. Each in their own way, have attempted to represent the phenomemon known as fire. There is no other natural phenomenon which so captivates the interest of so many. Man's struggle to control, extinguish, and prevent fires knows no geographical or political borders. These pictures vividly portray that struggle through the technology of photography.

The pictures in this book are the result of a national contest sponsored by Addison C. Getchell and Son, Inc. of Boston, Massachusetts. The entrees were judged by Henry M. Quinlan, President of Addison C. Getchell and Son, Inc., William F. Noonan, photographer for the Boston Fire Department and author of **Smoke Showin!,** and Glenn D. Usdin, Associate Publisher of Fire Engineering. No one seeing these pictures can help but be touched by the uniqueness and dramatization of fire photography.

Waiting for Help
Michael Schroeder
Port Huron, Michigan

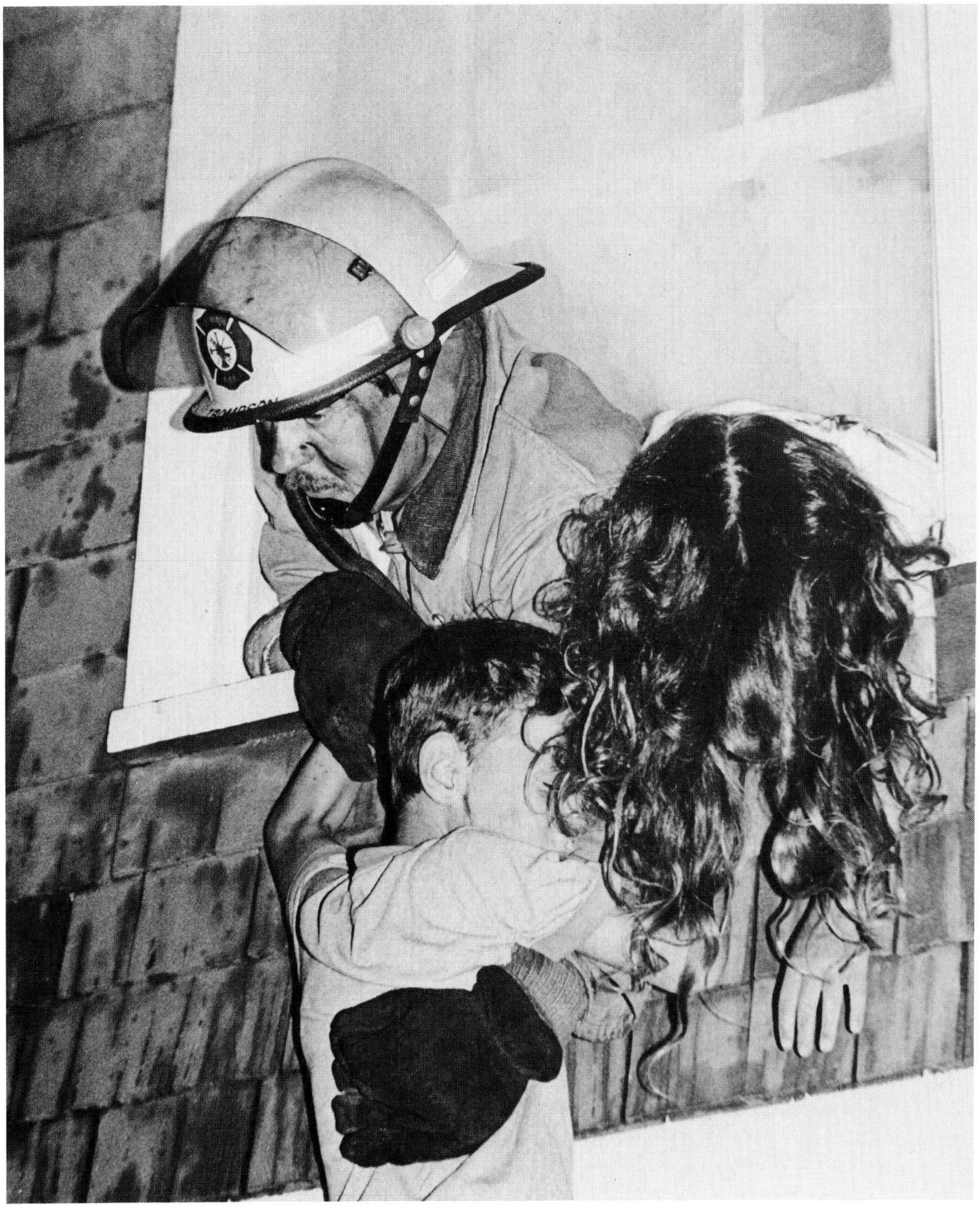

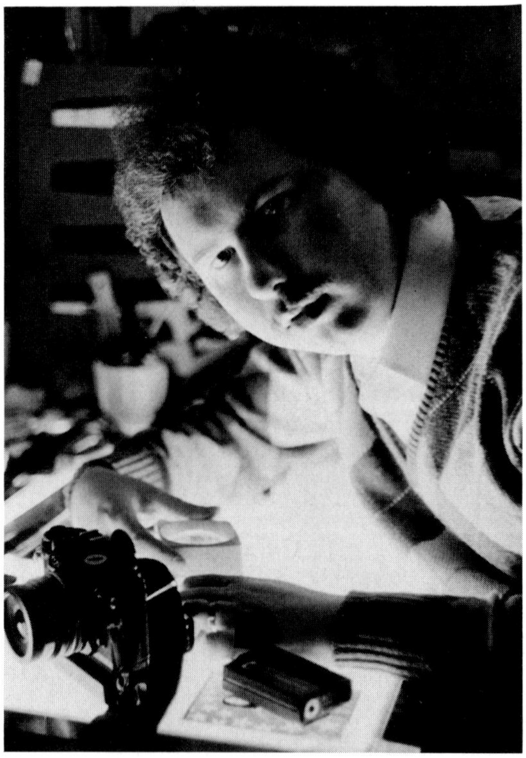

Michael Schroeder is a twenty-seven year old professional photographer who has owned his own studio for eight years. His studio assignments range from commercial to fashion. He also works as a freelance photojournalist for local newspapers and national magazines.

Mike became a member of the Photographic Society of America in 1981. That same year he was ranked second in the world in the Photojournalism Division. In 1984 he won the Young Photographer of the Year Award with his photo of a fireman rescuing a family's pet.

Mr. Schroeder has been covering fires since 1976. His interest in the fire field arises from the fact that his father is a Lieutenant on the City of Port Huron Fire Department.

Mike has won numerous awards for his achievements in spot-news photography.

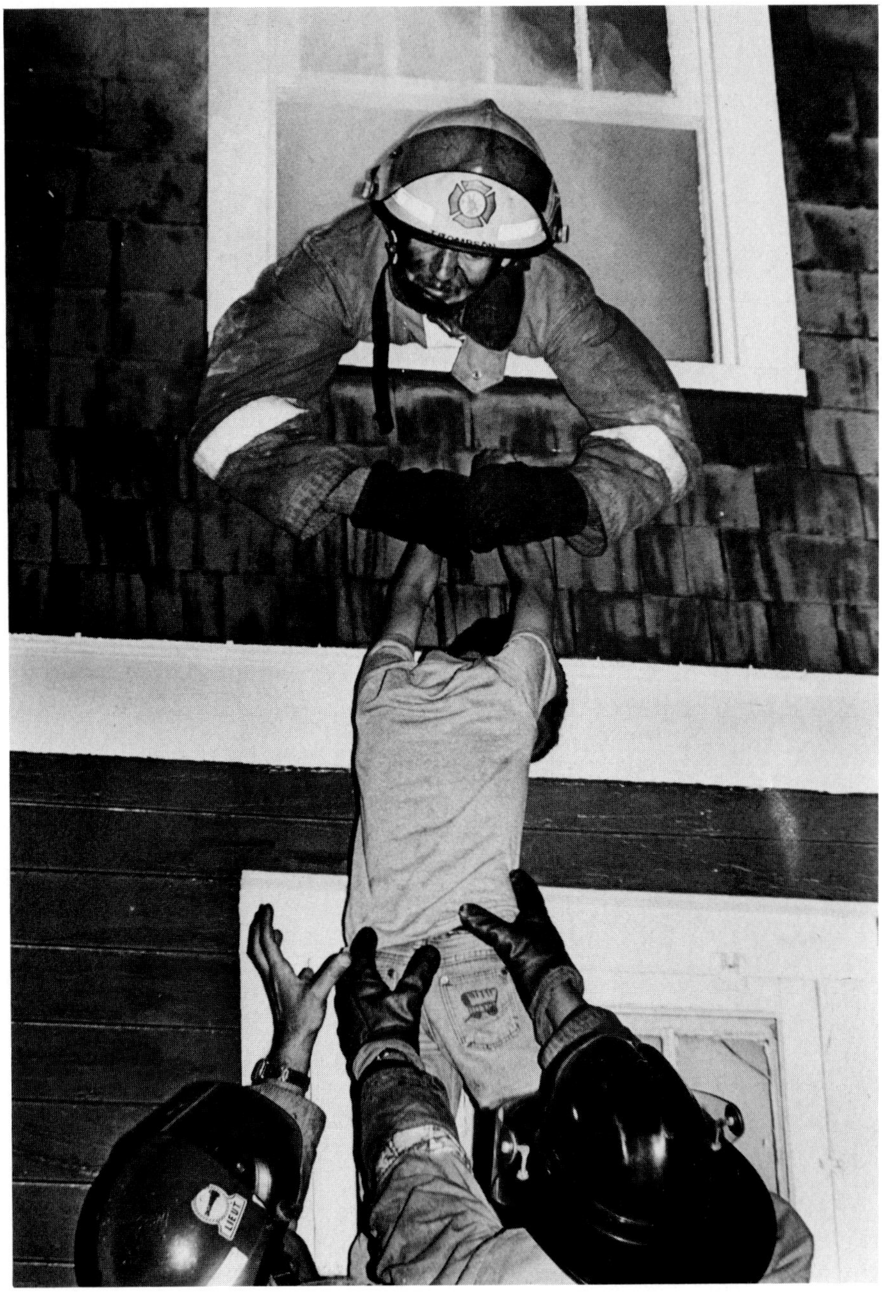

Second Story Rescue
Michael Schroeder
Port Huron, Michigan

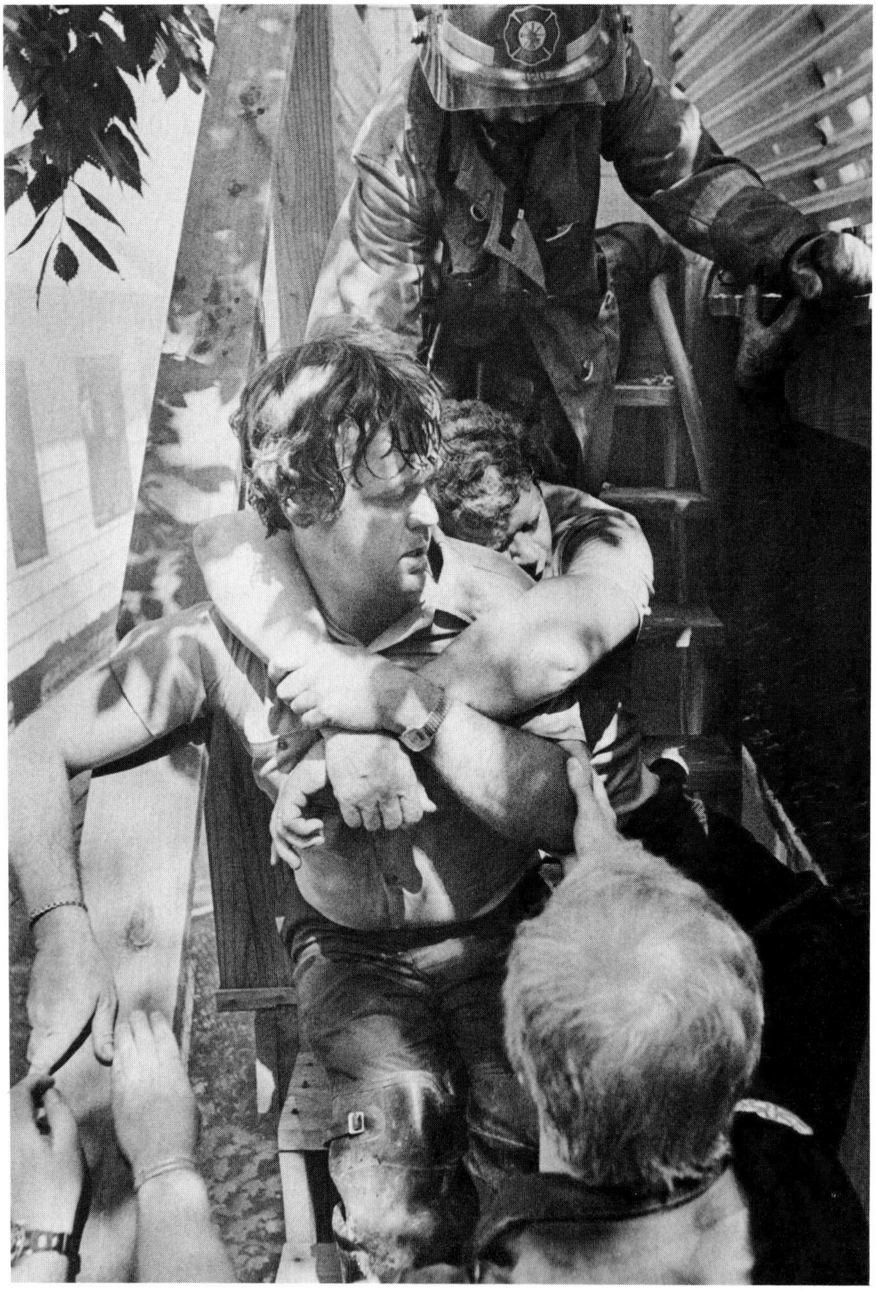

Firefighter Carried to Safety
Michael Schroeder
Port Huron, Michigan

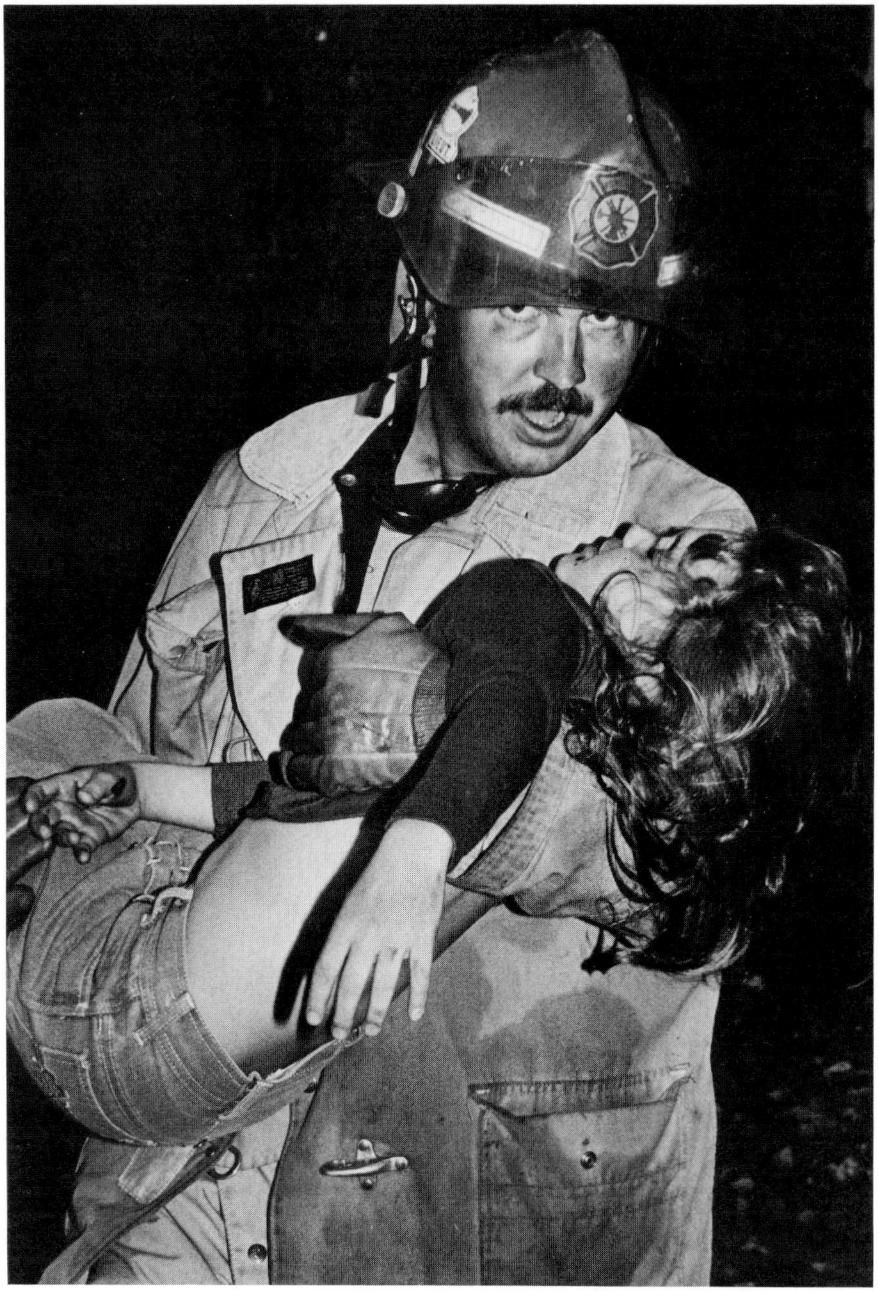

House Fire Rescue
Michael Schroeder
Port Huron, Michigan

Firefighter Being Rescued by Fellow Firefighters
Howard Meile, III
Baltimore, Maryland

2 ND **PRIZE**

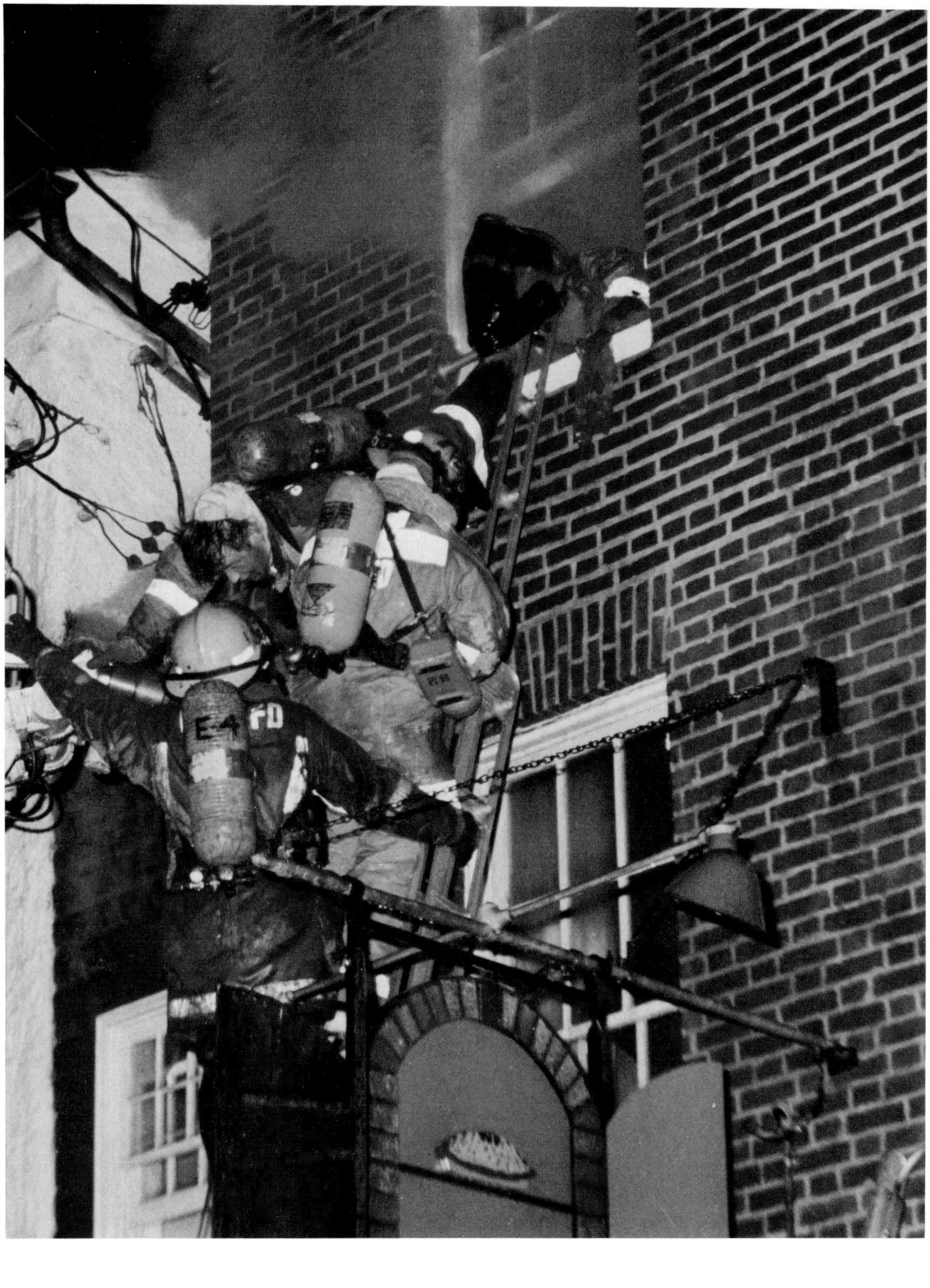

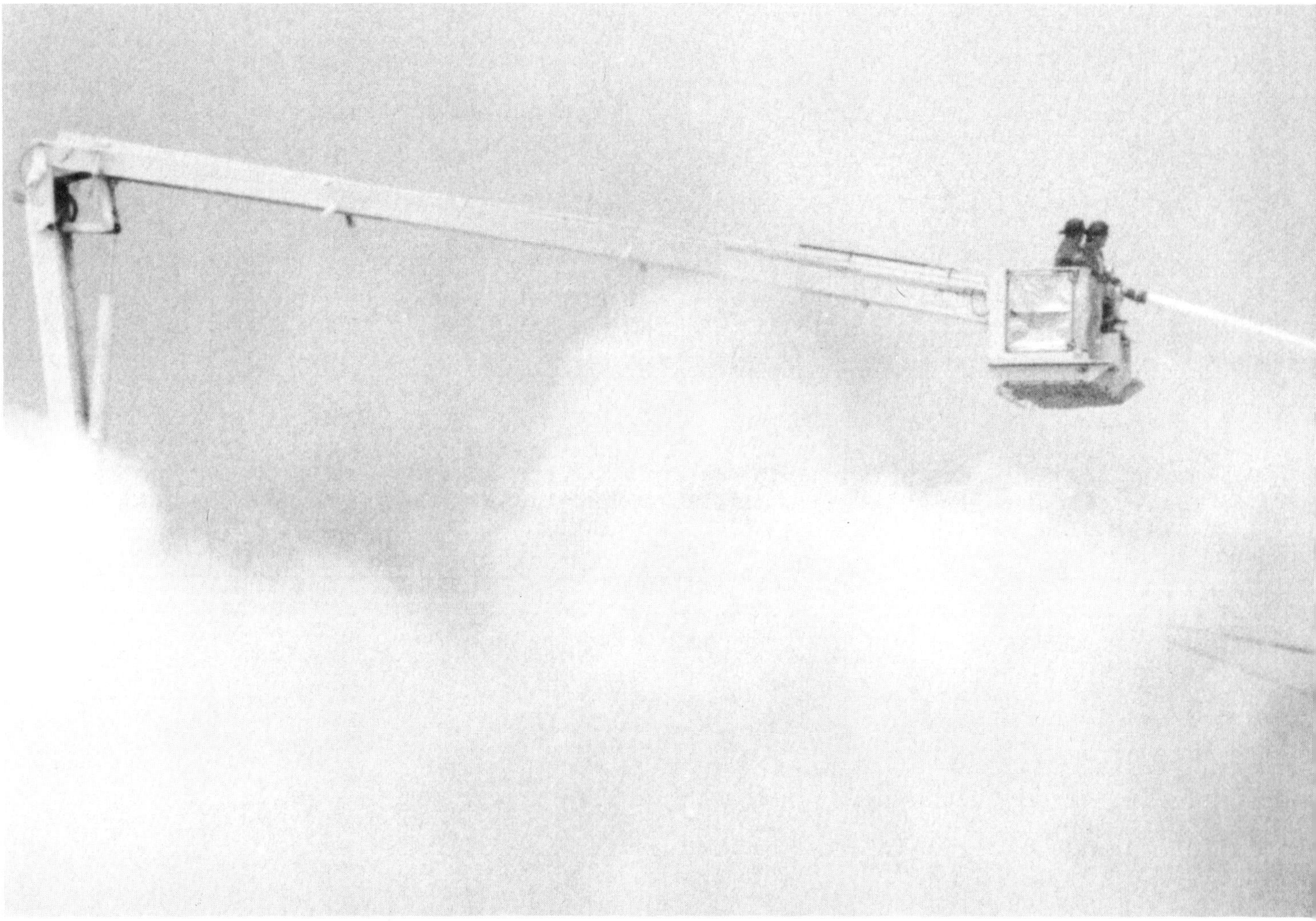

Snorkel Emerging from Smoke
Howard Meile, III
Baltimore, Maryland

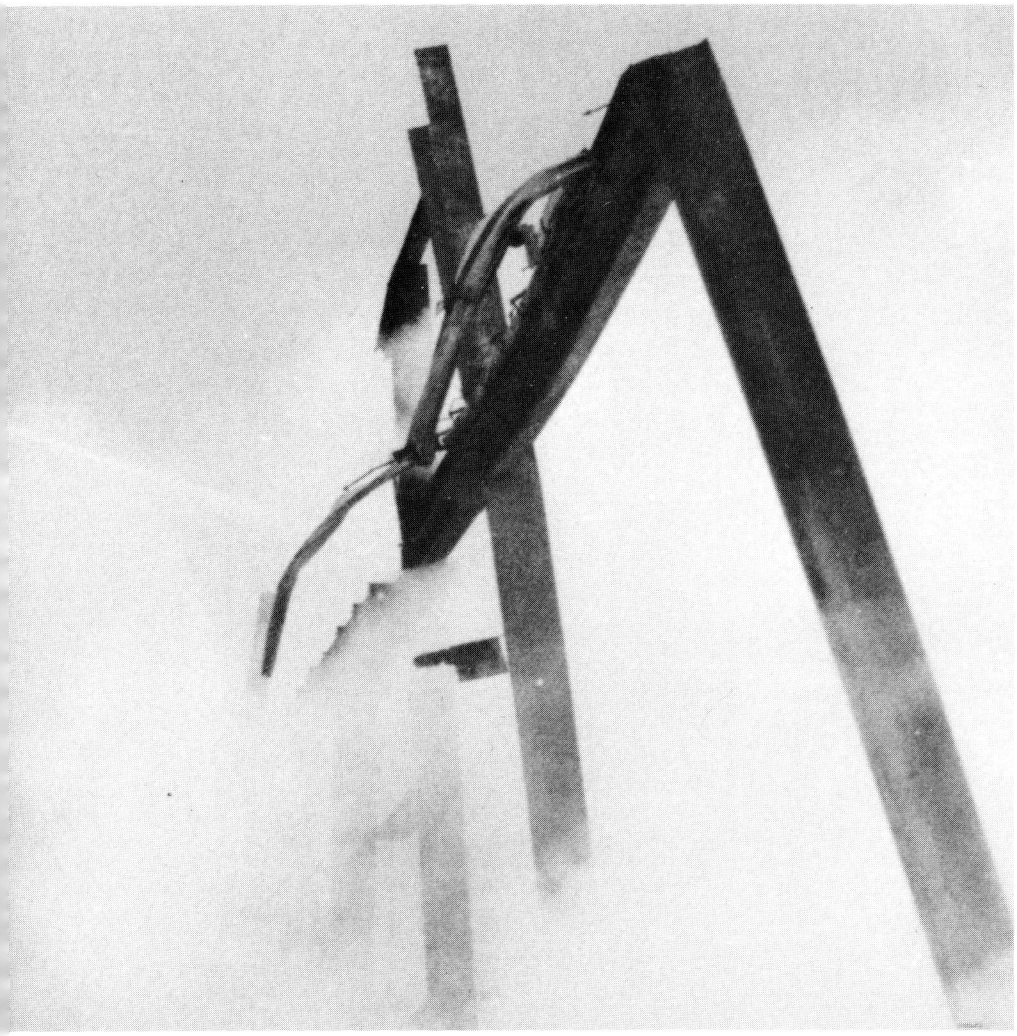

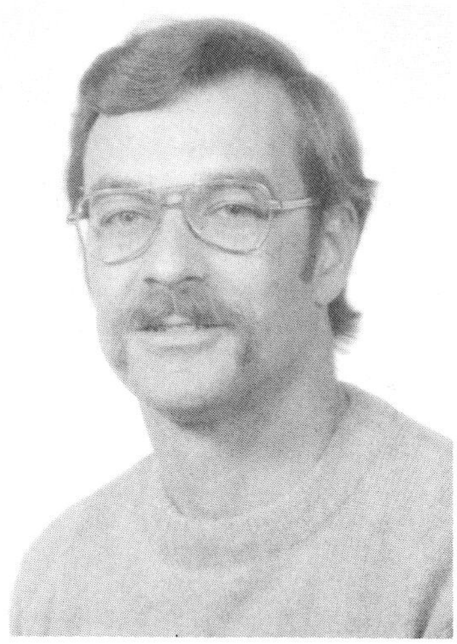

Howard Meile, III is 33 years old and has lived in Baltimore all his life. He has been photographing fires and fire apparatus since he was 14 years old. He is an Auxiliary Firefighter with the Baltimore City Fire Department, running out of Truck Company 10 (the busiest running truck in the city of Baltimore).

He is a member of the International Fire Photographers Association, and of the Box 414 Association, a fire buff organization in Baltimore City.

He has won several awards and prizes with his fire photographs. Some of his photographs have been published in Firehouse Magazine and The Baltimore Firefighter, a paper for the firefighters of Baltimore City.

Aftermath: Triangle Refineries Five

M. Jack Luedke

Jacksonville, Florida

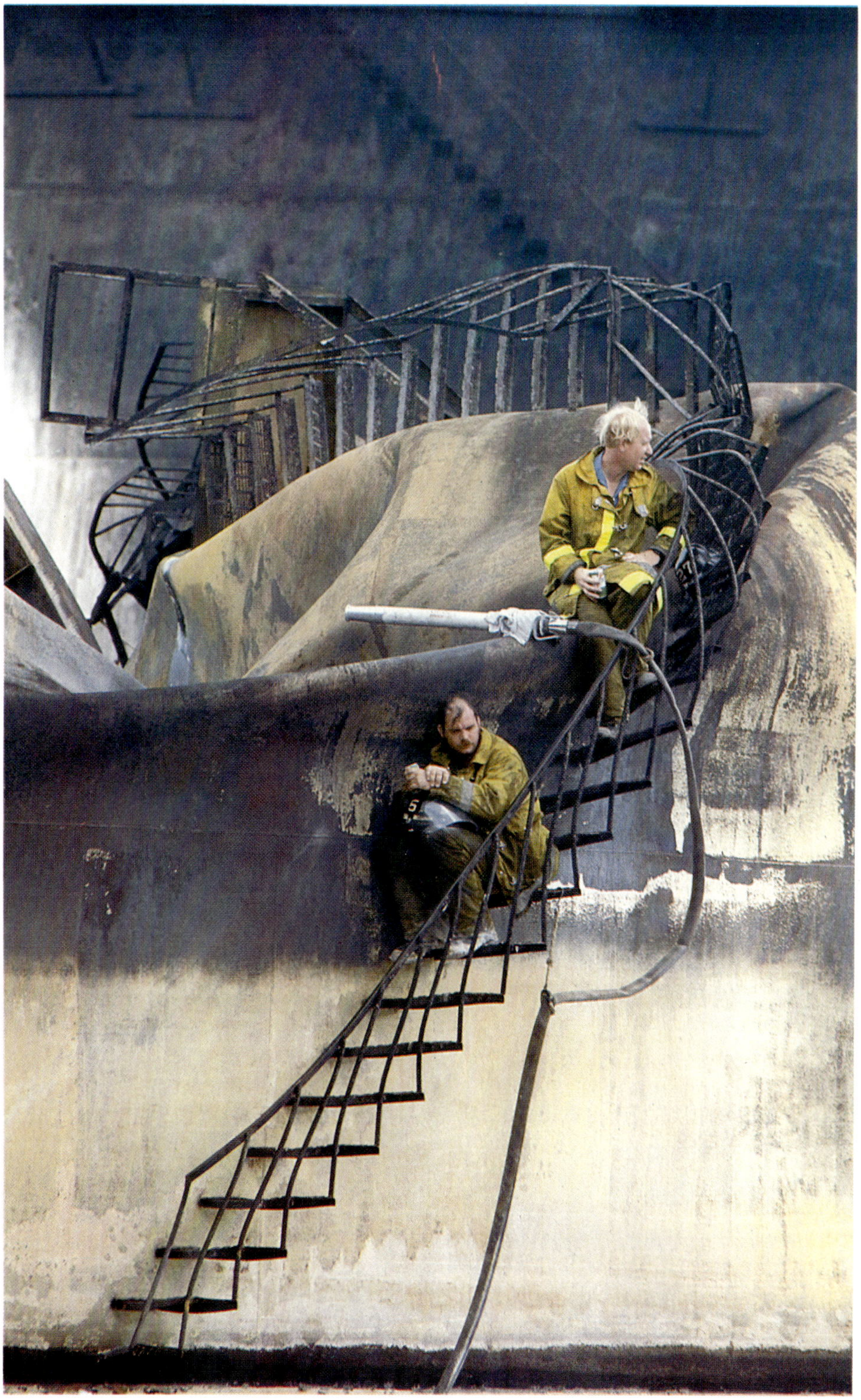

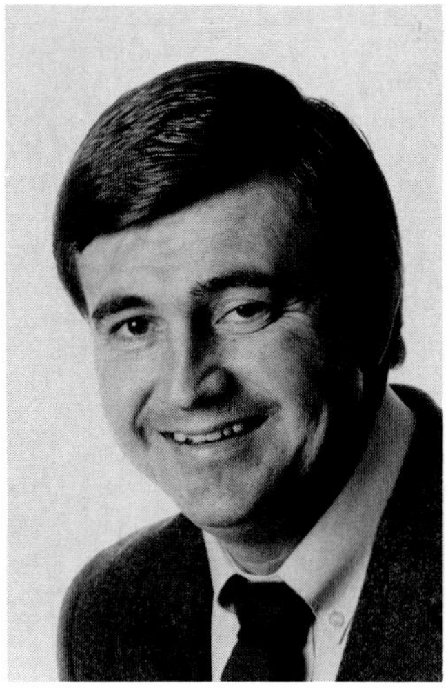

Melvin Jack Luedke was born January 17, 1953, in Tucson, Arizona. He grew up in Tucson, Ariz.; Roswell, N.M.; and Goose Bay, Labrador, Can.

Jack, as he is called by his friends, graduated from Rincon High School in Tucson, in 1971 in the top 15 percent of a class of over 600.

In 1971-72 Jack attended the University of Arizona in Tucson where he studied anthropology.

In 1973 he joined the U.S. Navy and successfully completed his training at the U.S. Navy Photographic Intelligence School in Denver, Colorado. He received an Honorable Discharge in 1977 but remains an Intelligence Specialist in the U.S. Naval Reserve.

Jack received his Associate of Arts degree in photography and graphic arts at the University of North Florida in 1982.

Jack did an internship for college credit with the Florida Times-Union in 1979 and was the photo editor, editorial cartoonist, and illustrator for the University of North Florida Spinnaker Newspaper in 1979-80. He also worked as a teaching assistant and darkroom technician for the university during that time.

Jack gained his initial newspaper experience with Clay Today newspaper in Orange Park, Fla., from 1980 to 1983.

He has been employed as a staff photographer with The Florida Times-Union since April 1983.

In 1984 Jack placed second in the news category of the NPPA's National Clip Contest for the month of August.

Luedke lives in Jacksonville with wife, Gayle, and their two children, Sarah, 4, and Nikolas, 6 months. Gayle is a supervisor/bookkeeper for Pantry Pride Inc. in Jacksonville.

Luedke has many interests but his passions are photography, his family and his friends.

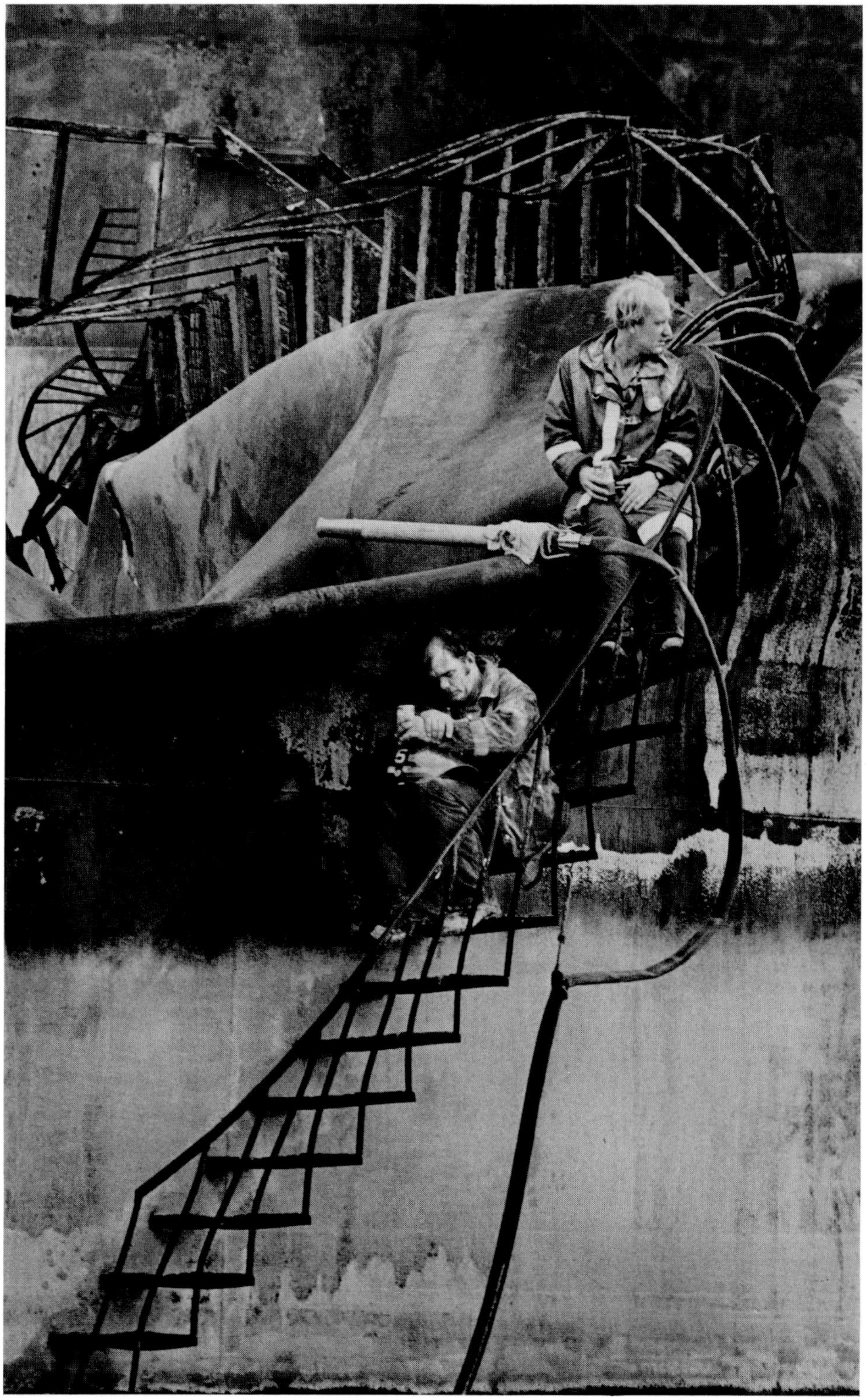

Ordeal by Fire: Triangle Refineries Fire
M. Jack Luedke
Jacksonville, Florida

Honor Award Photographs

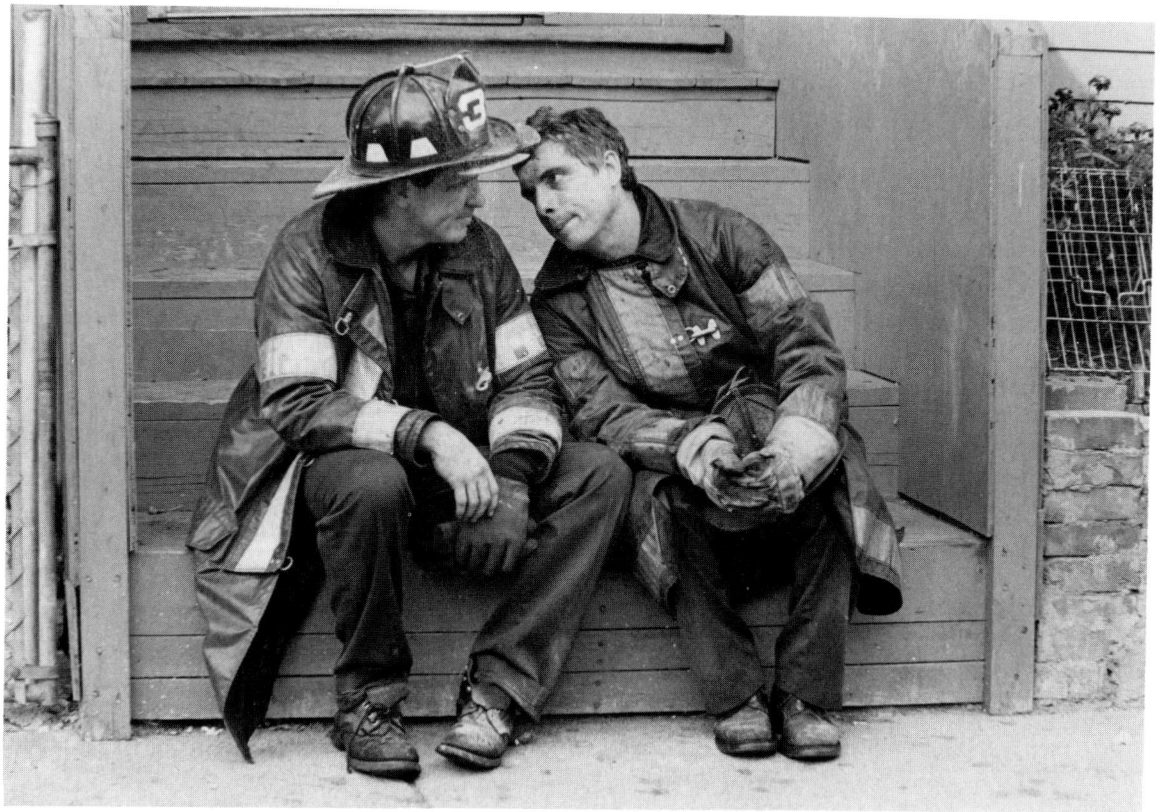

Untitled
James Marabello
Fitchburg, Massachusetts

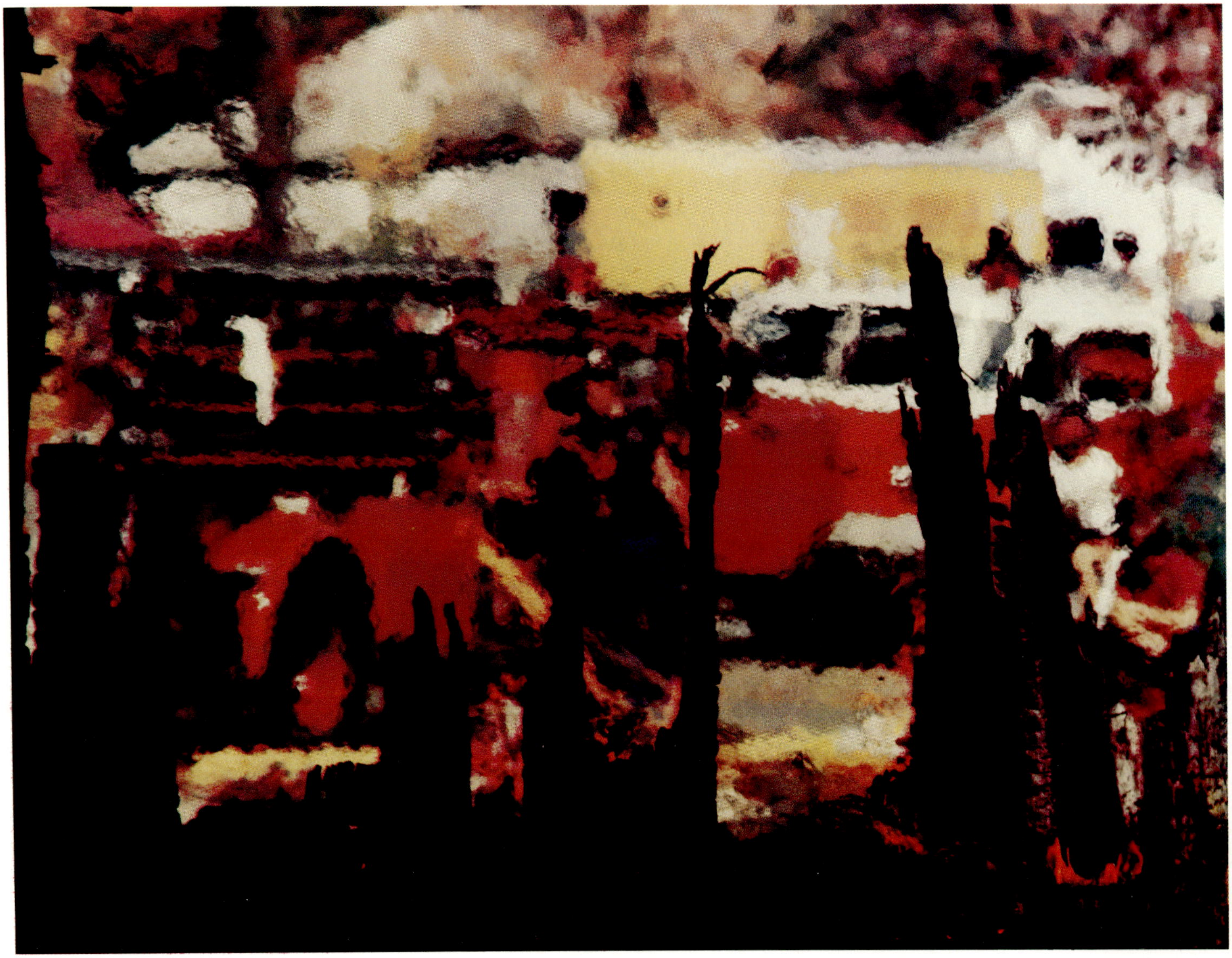

Fire Abstract
Glen E. Ellman
Dallas, Pennsylvania

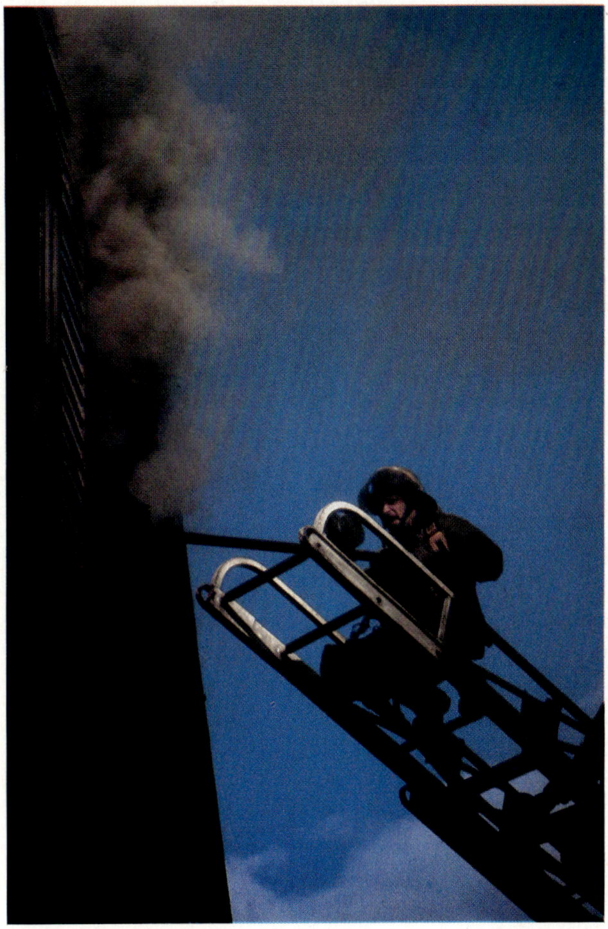

John A. Cannon
Newington, New Hampshire

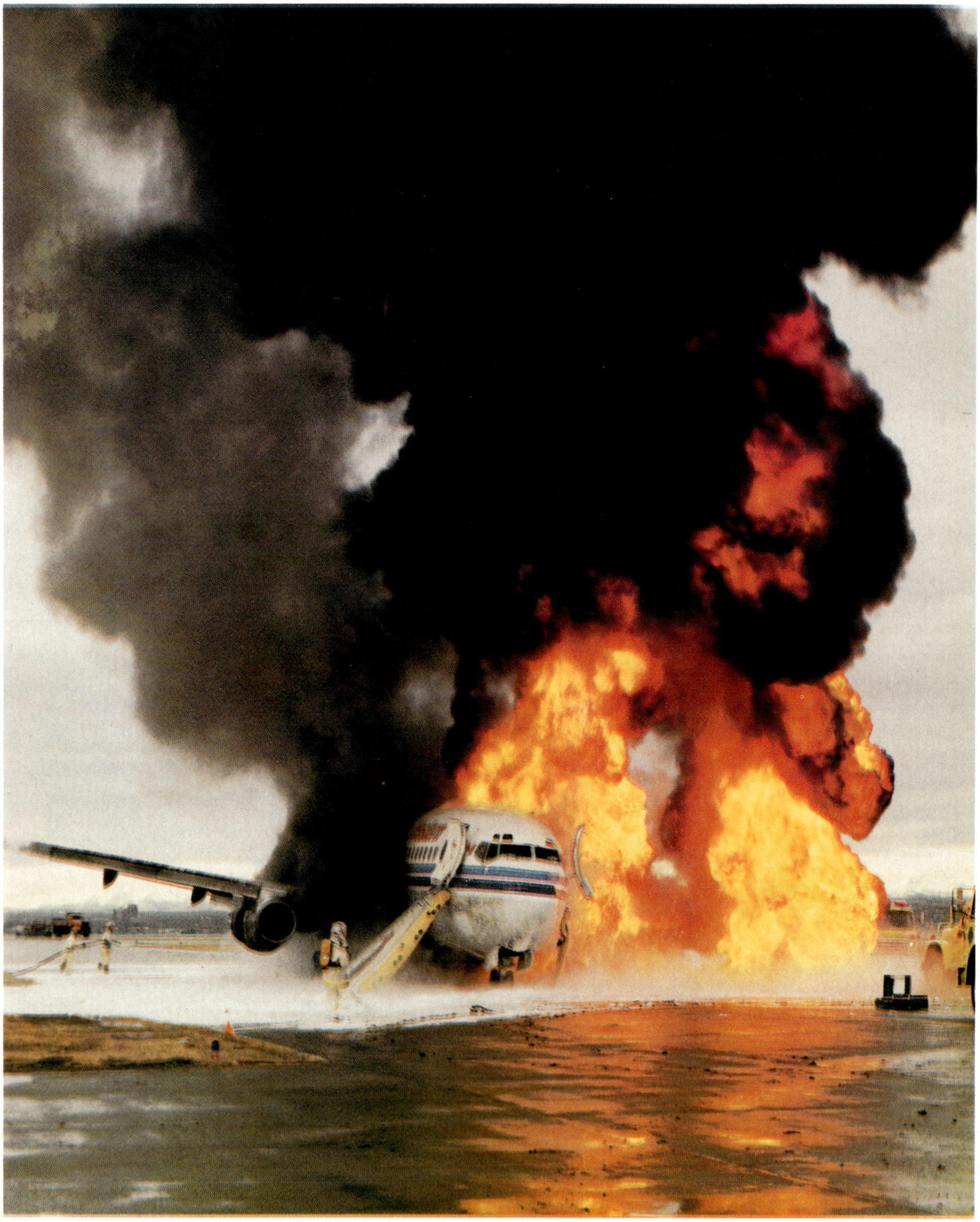

A Pilot's Nightmare
O.J. Tveter
Calgary, Alberta, Canada

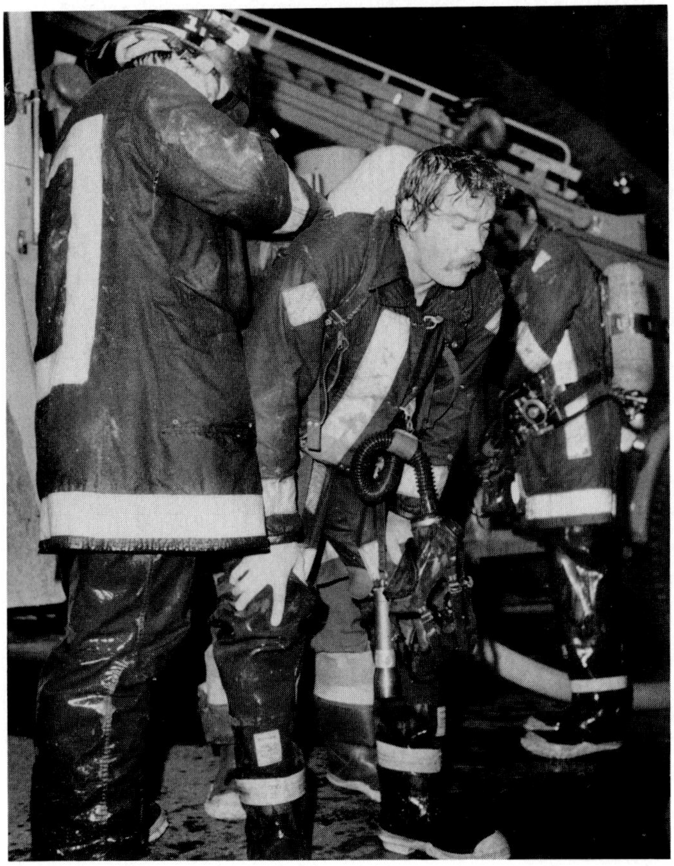

Get Air Tank Changed
Mike Waters
DeWitt, New York

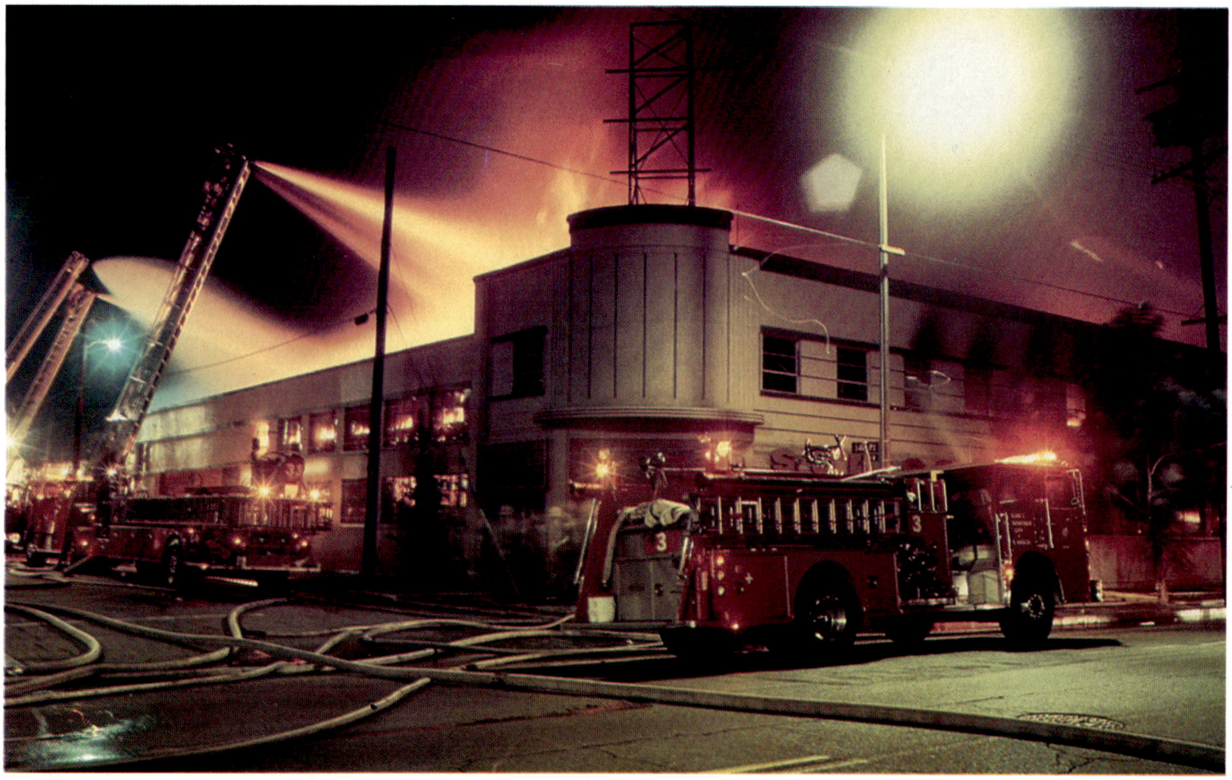

City Lights
Keith D.Cullom
Goleta, California

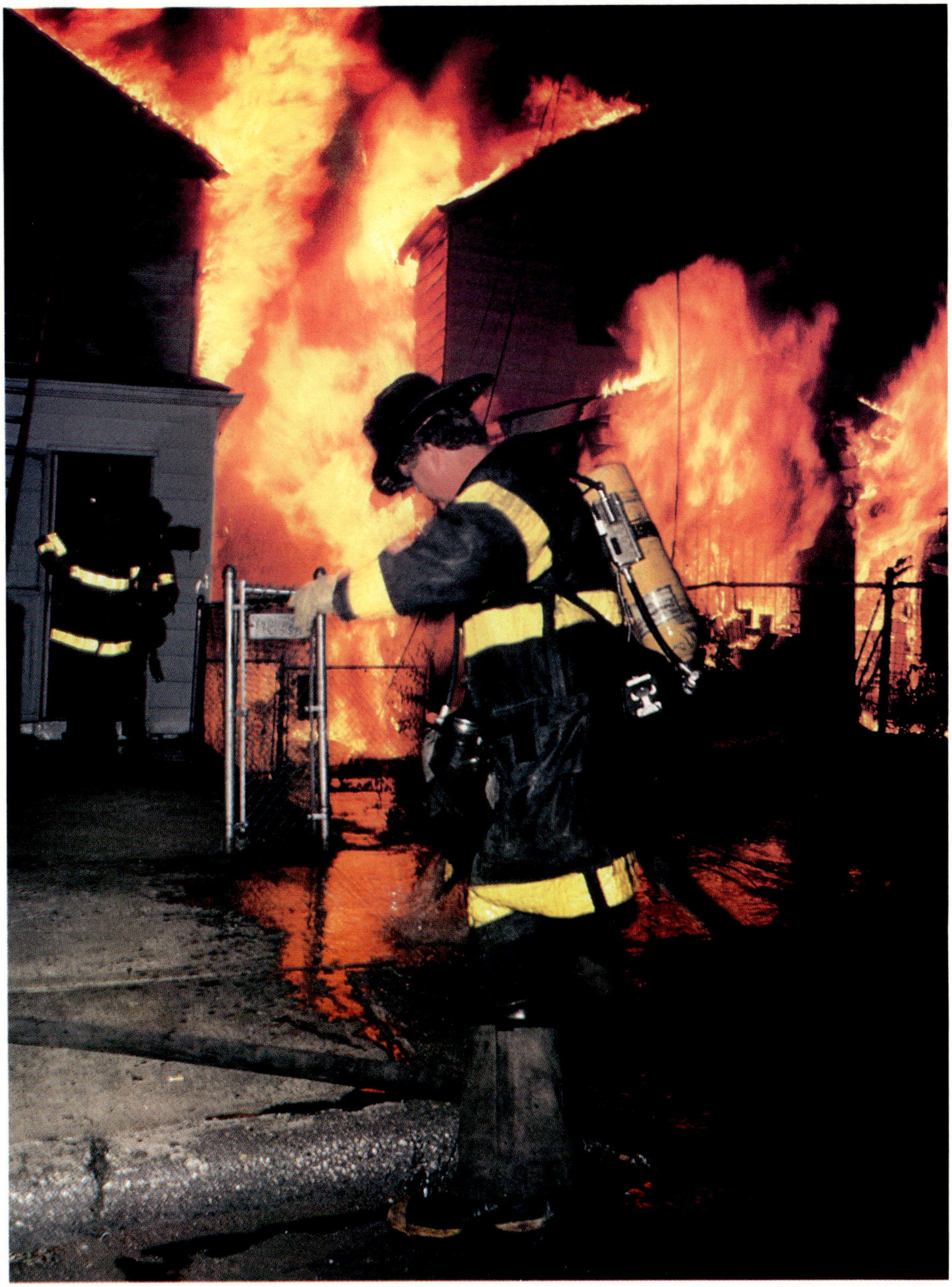

Move Into The Fire
Keith D. Cullom
Goleta, California

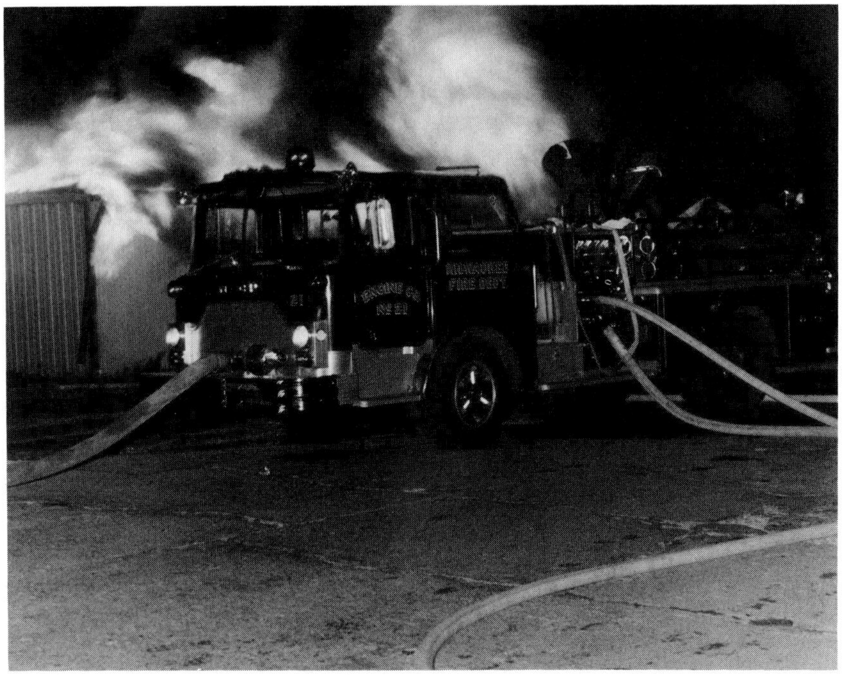

Setting Up Deluge Set
Mark Hoeller
Milwaukee, Wisconsin

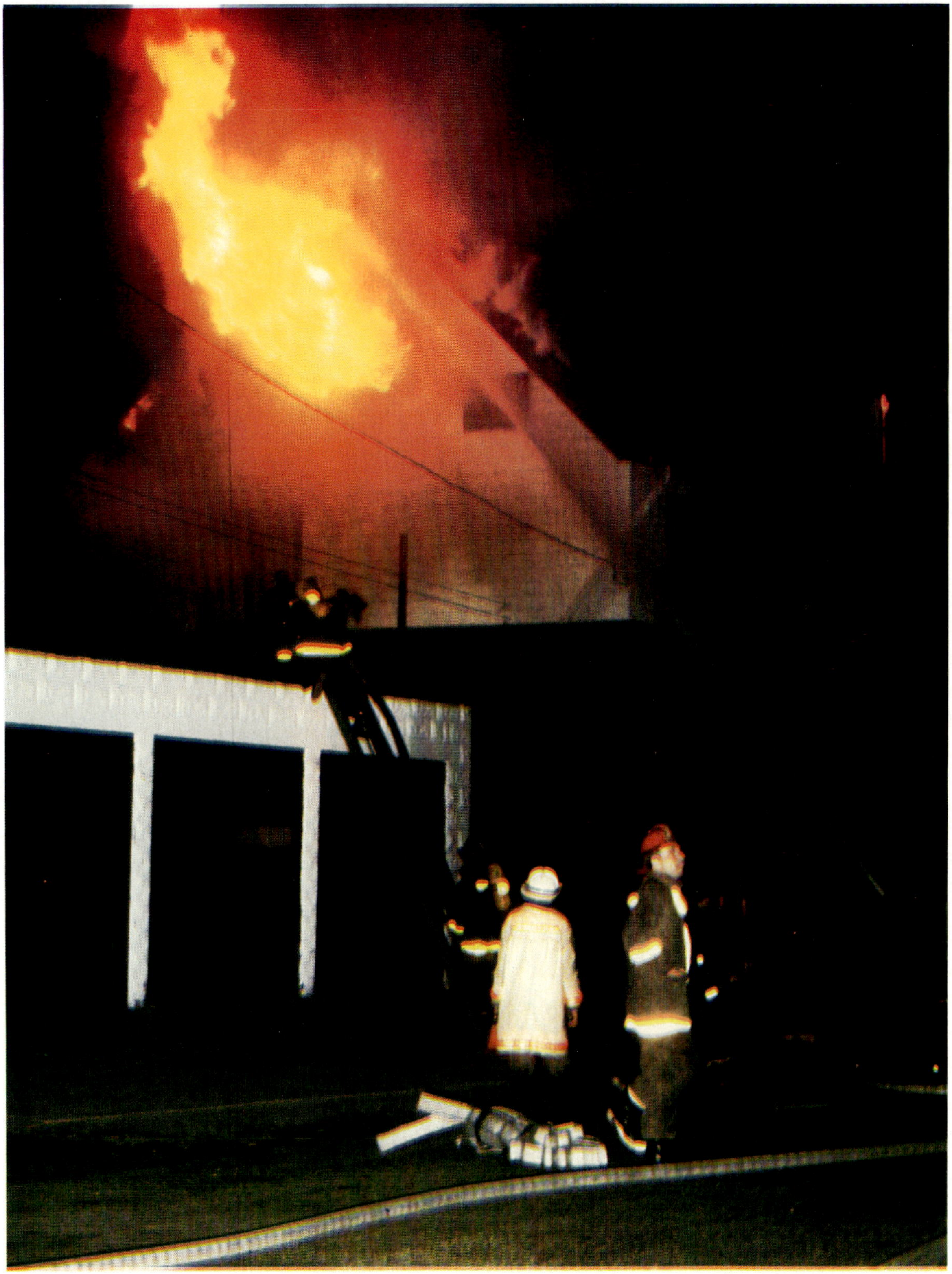

Blow Out
Mark Hoeller
Milwaukee, Wisconsin

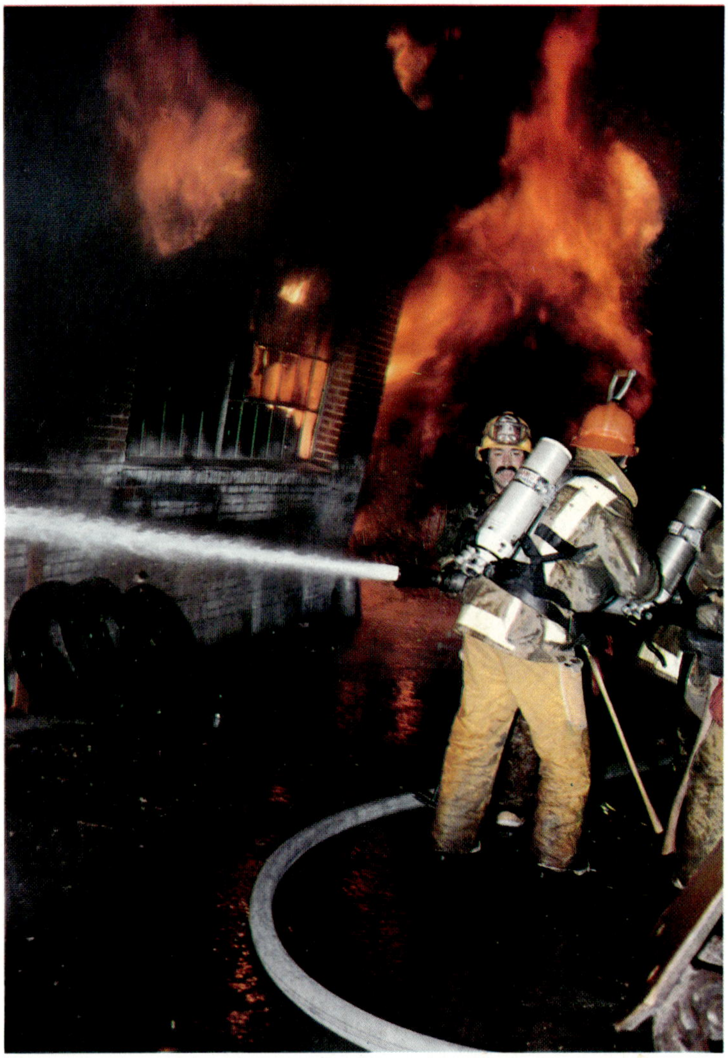

Get Out
Alan Simmons
Sun Valley, California

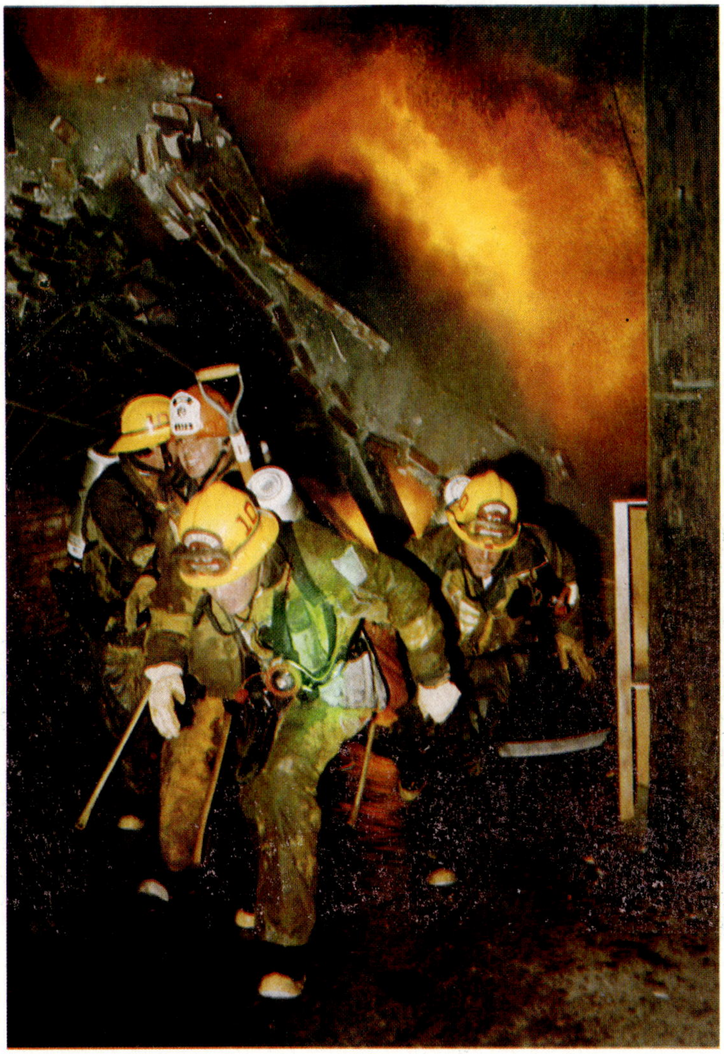

On The Run
Alan Simmons
Sun Valley, California

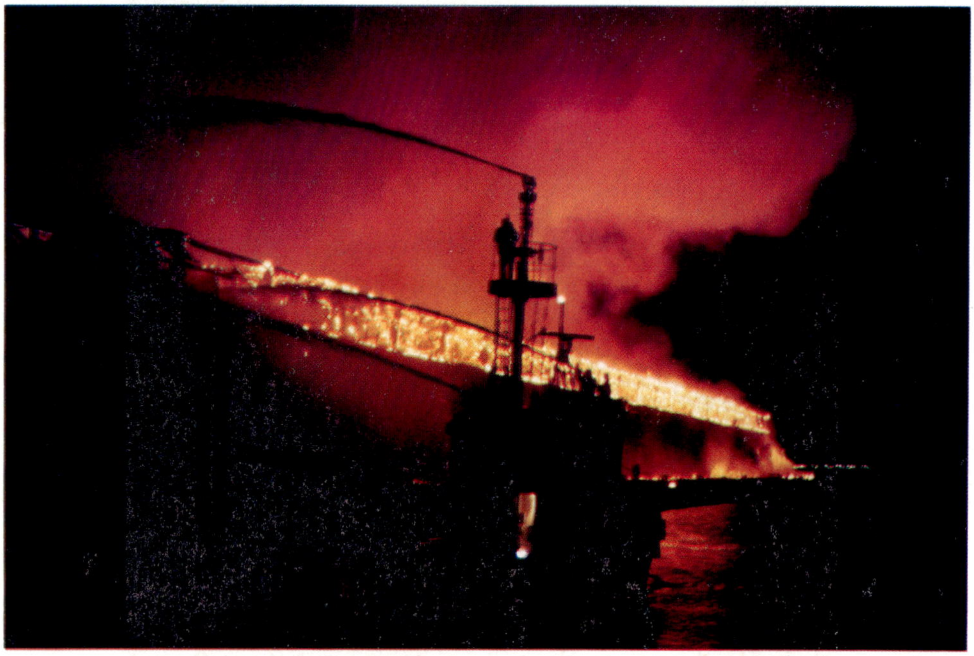

5th Alarm - Pier 30-32
Michael S. Newman
San Francisco, California

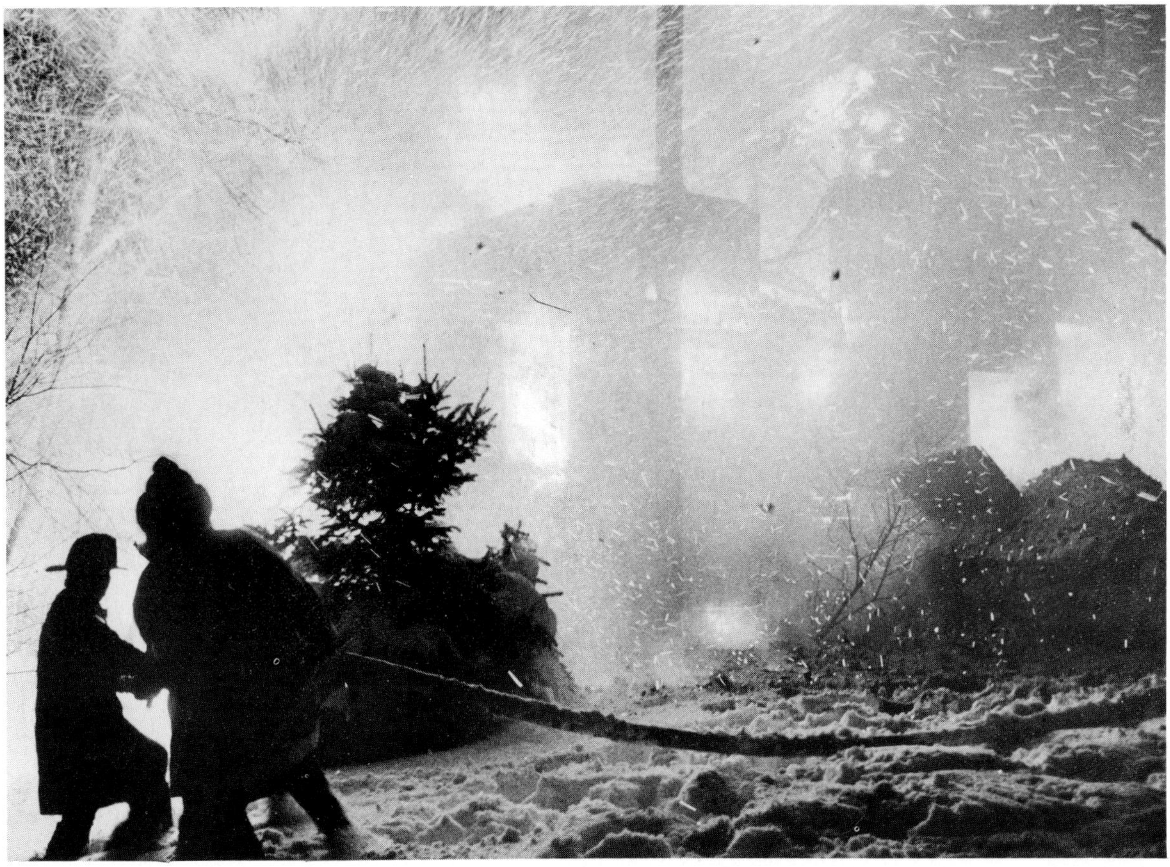

Firefighters Move Back as a Building Collapses
Showering Them With Embers.
Bob Stella
Braintree, Massachusetts

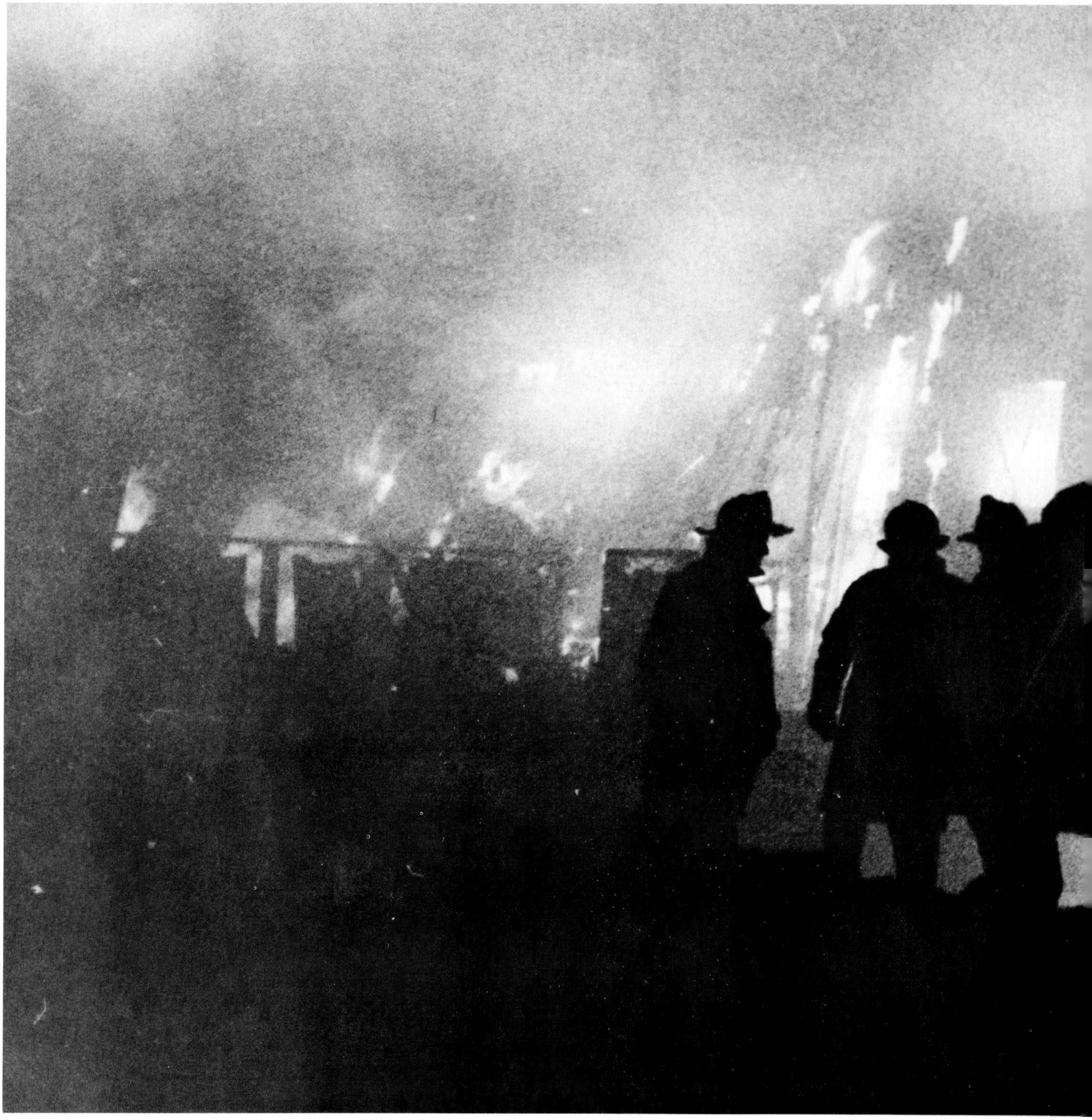

Fire on a Pier Behind North Station
Bob Stella
Braintree, Massachusetts

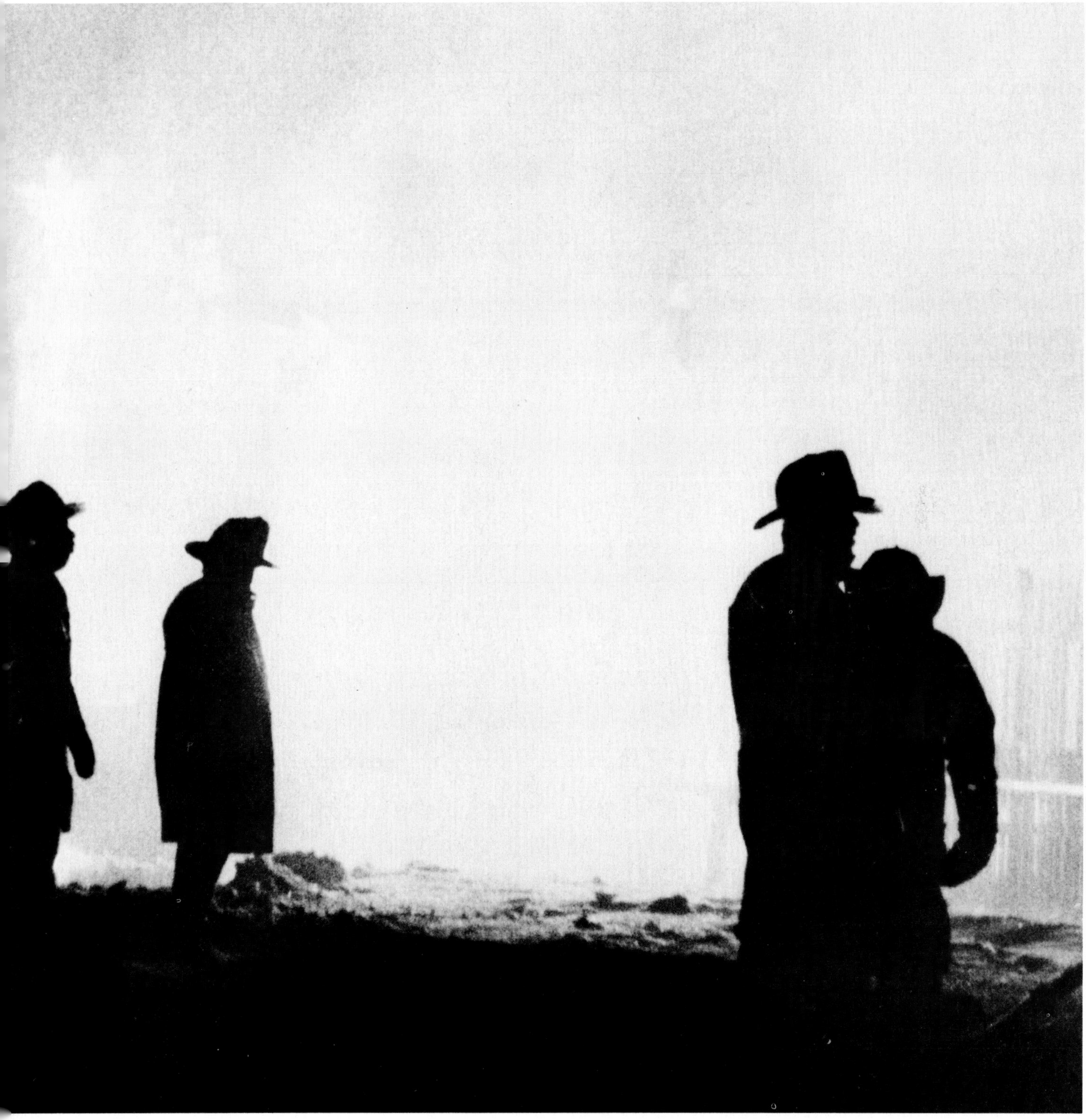

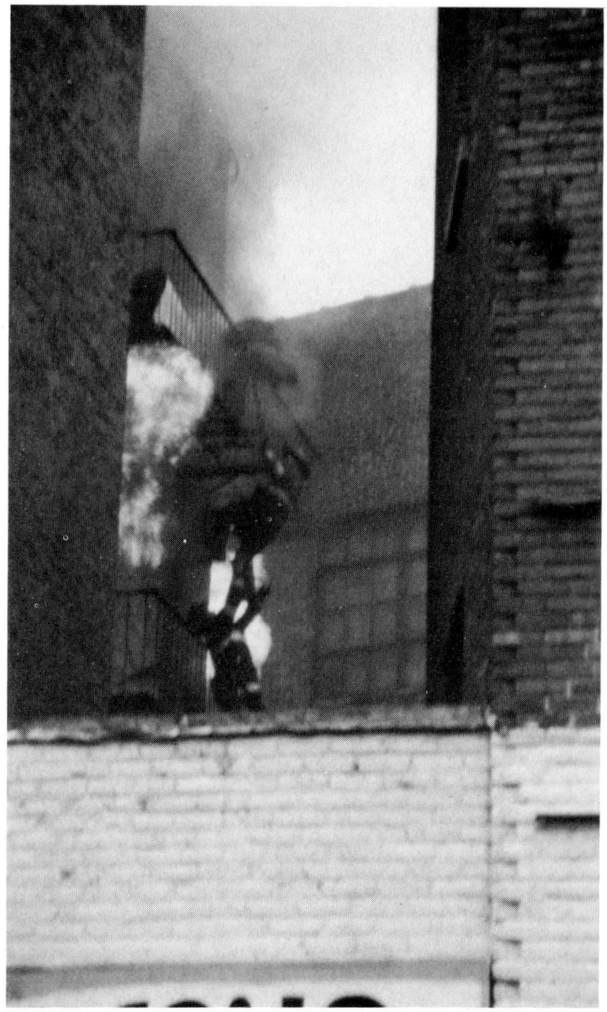

Untitled
Warren Fuchs
Brooklyn, New York

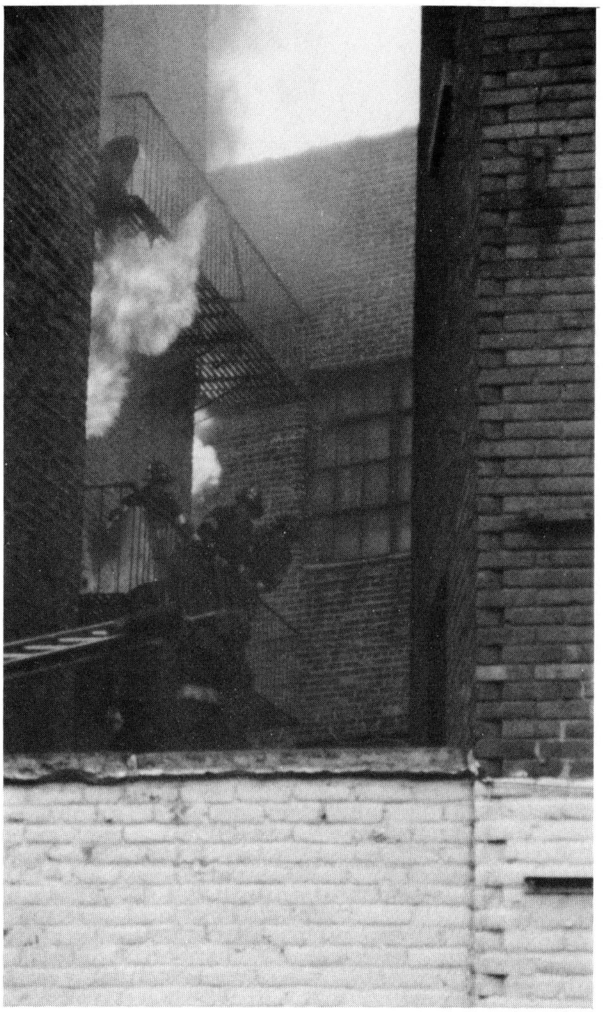

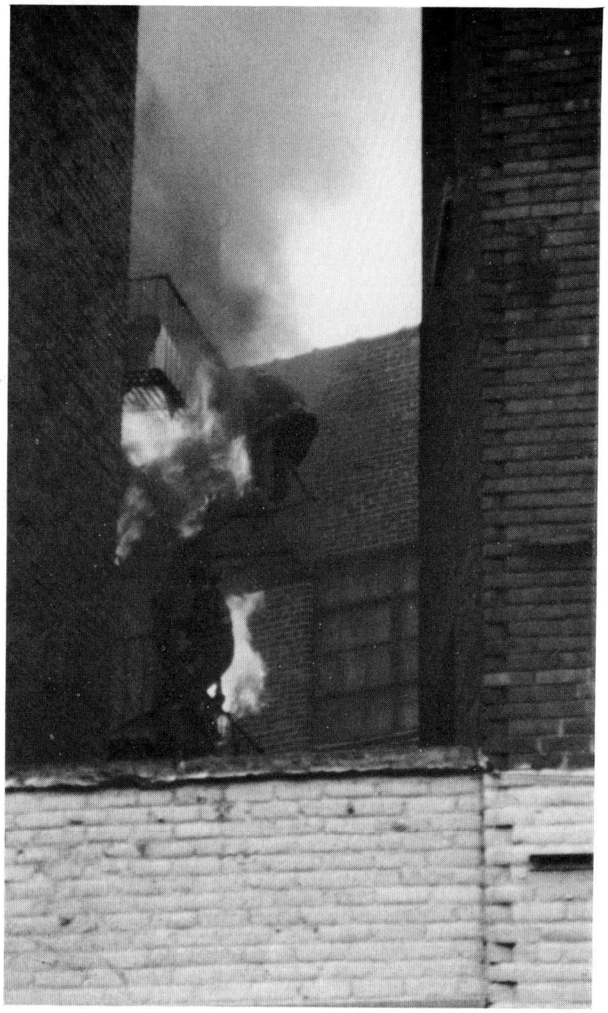

Honorable Mention Photographs

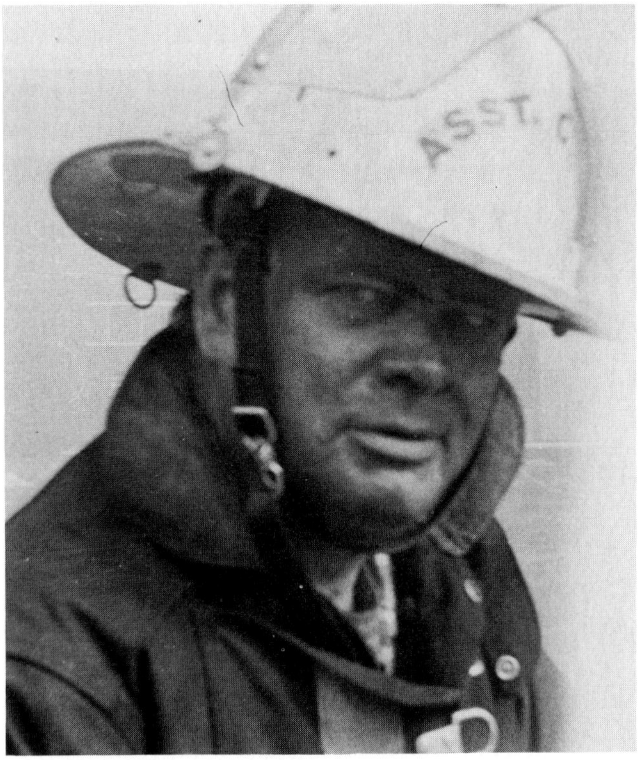

The Volunteer
Jerry Johnson
Chandler, Oklahoma

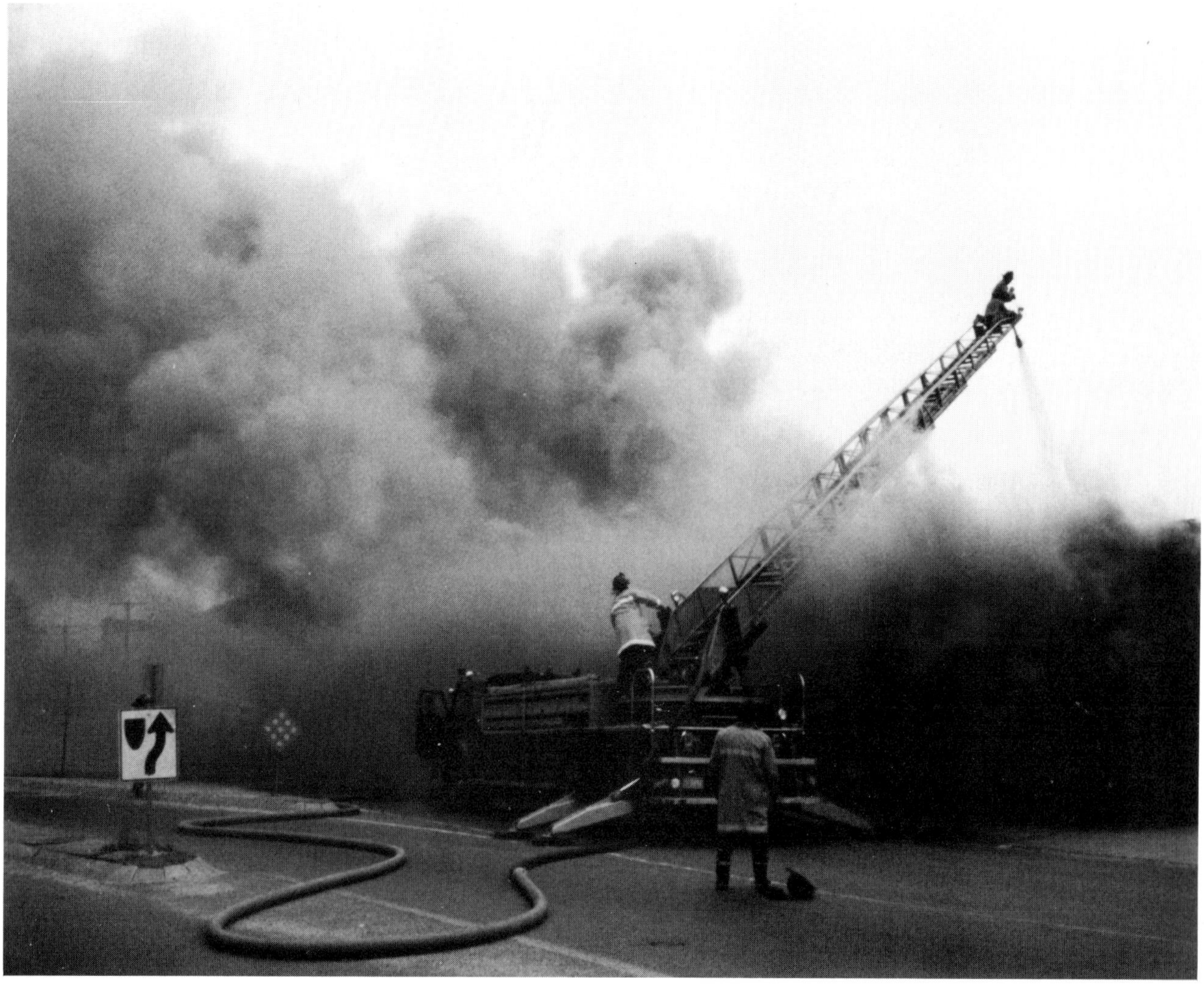

John C. DeForest
Pittsfield, Massachusetts

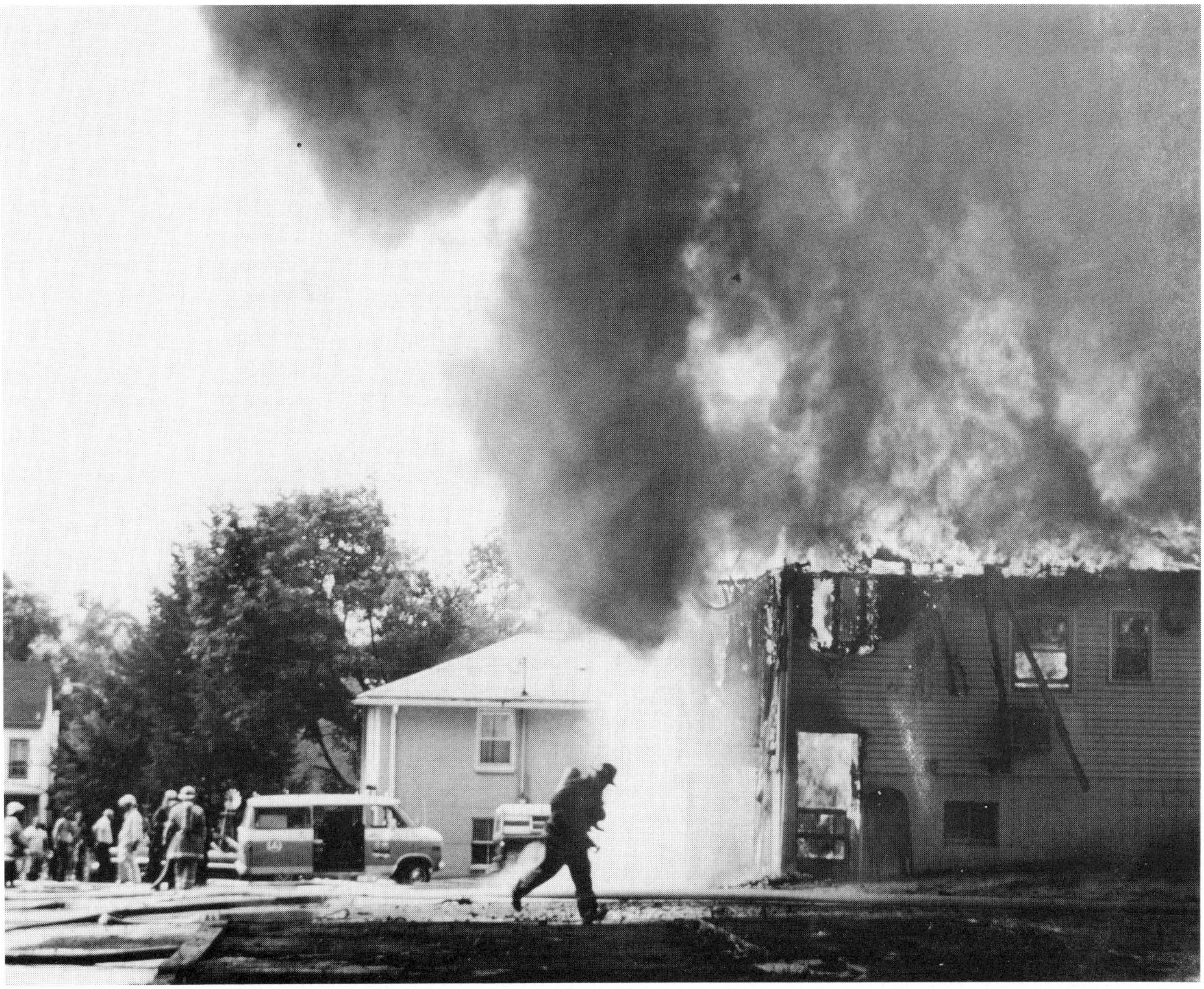

The Scorcher
Paul D. Renn, Jr.
Middletown, Pennsylvania

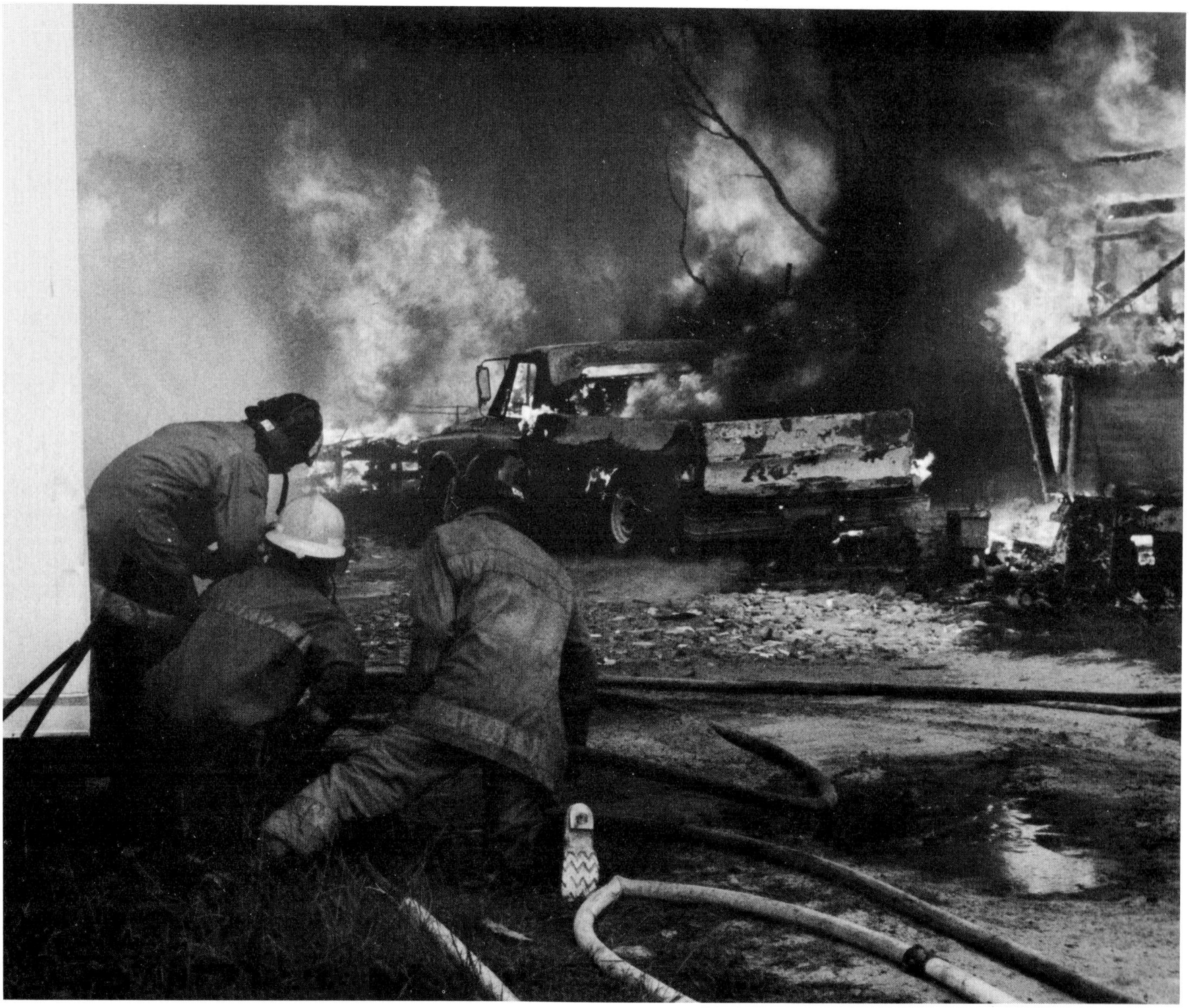

In Heat
Jeffrey Phyall
Lakeland, Florida

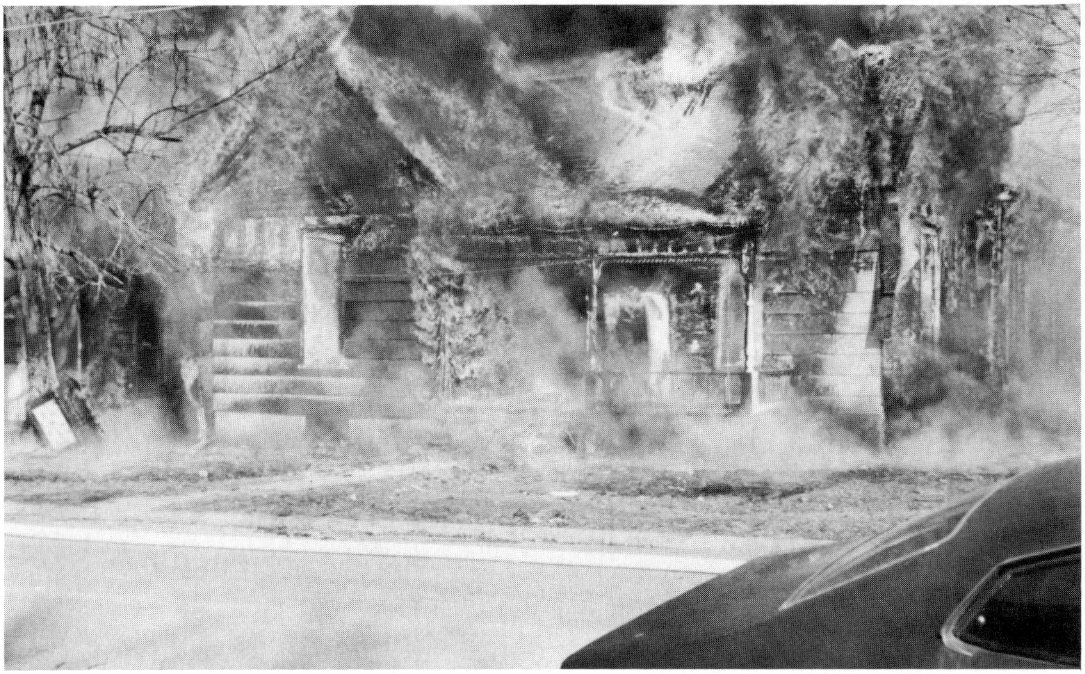

Arson Fire at Lyons & Morgans in Marshall, MO
Rhonda Herron
Marshall, Missouri

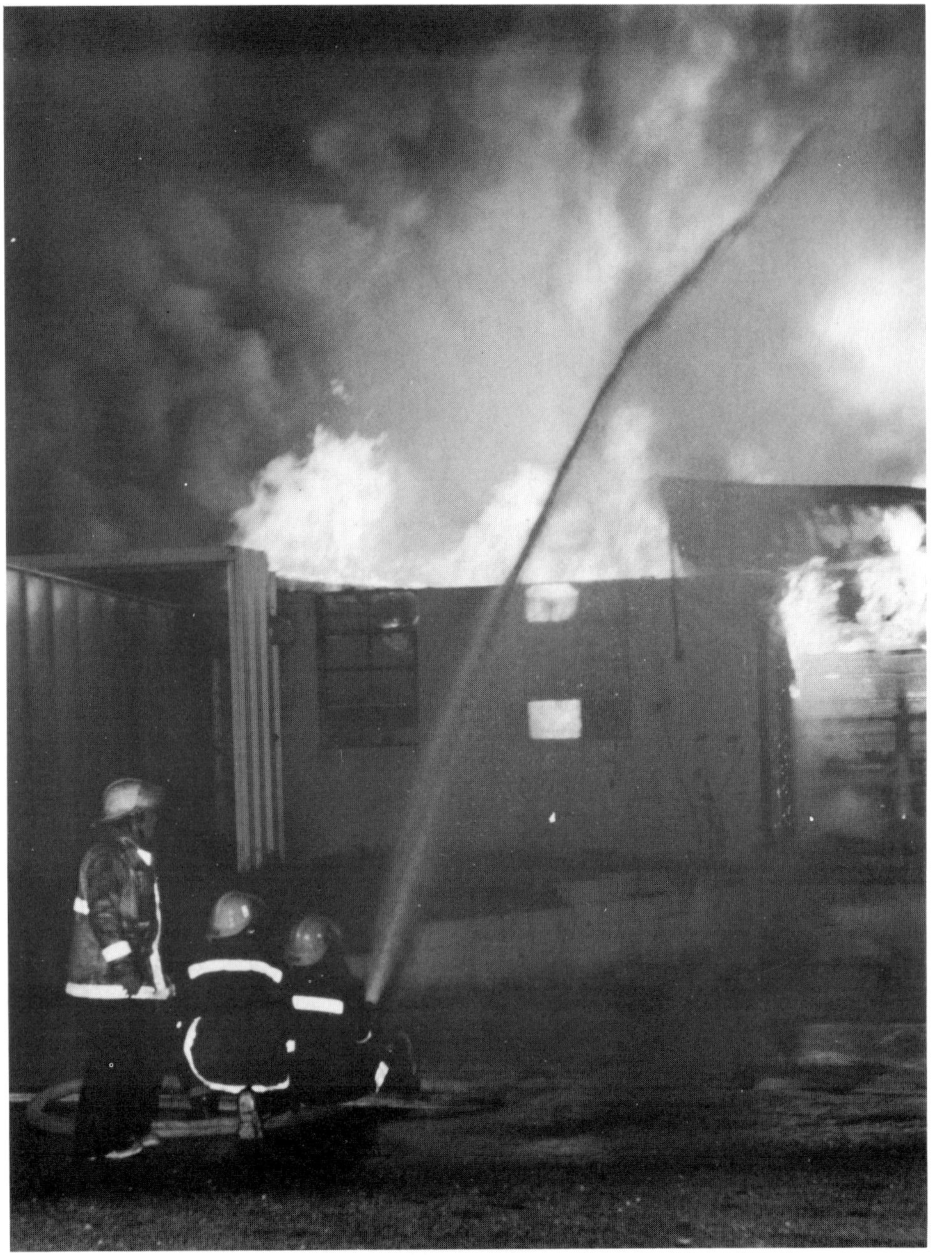

Raging Fire Claims $2 Million Dollar Loss
Dreama Floersch
Milan, Tennessee

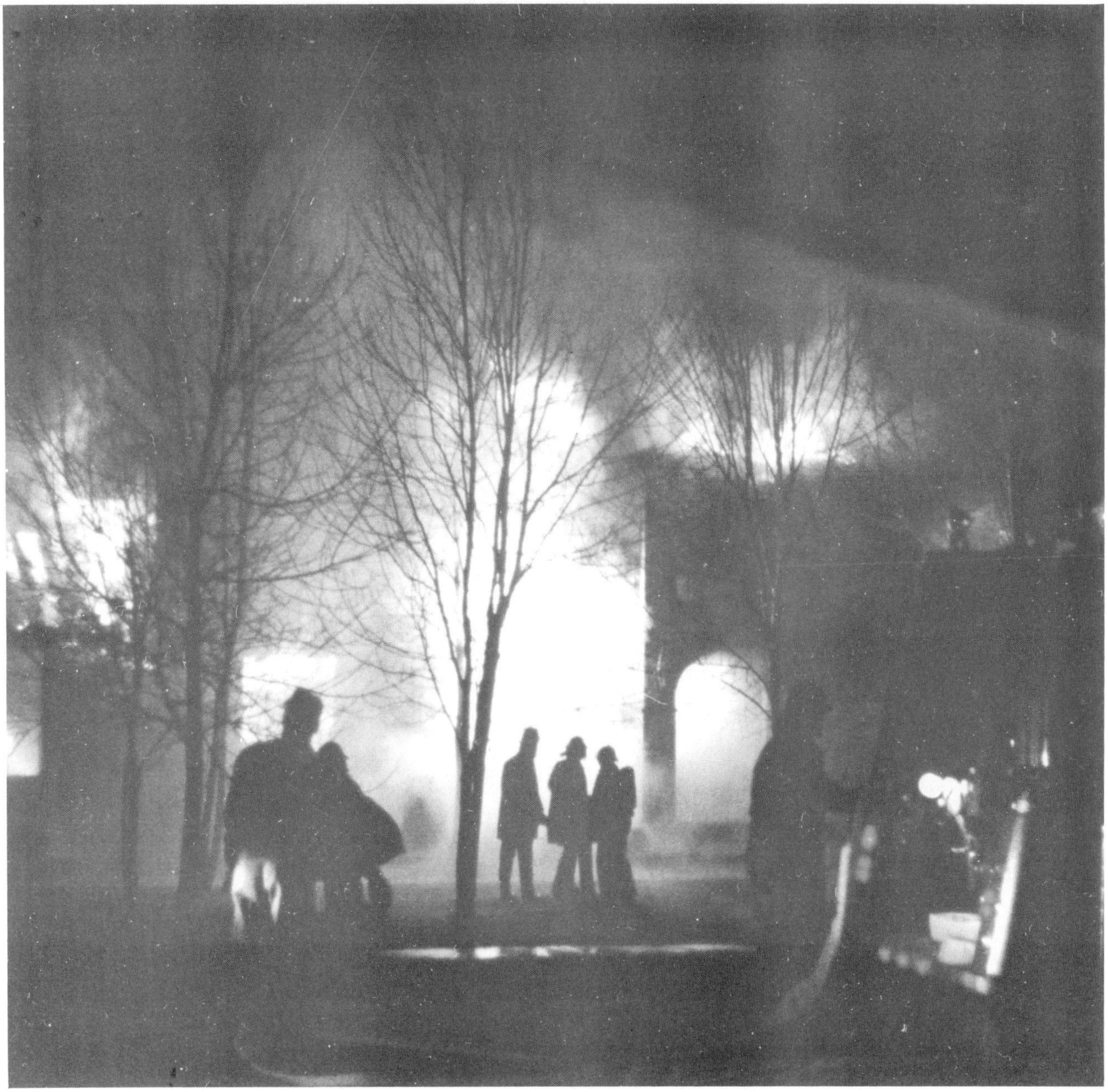

Fully Involved
Mark A. Dettmer
Getzville, New York

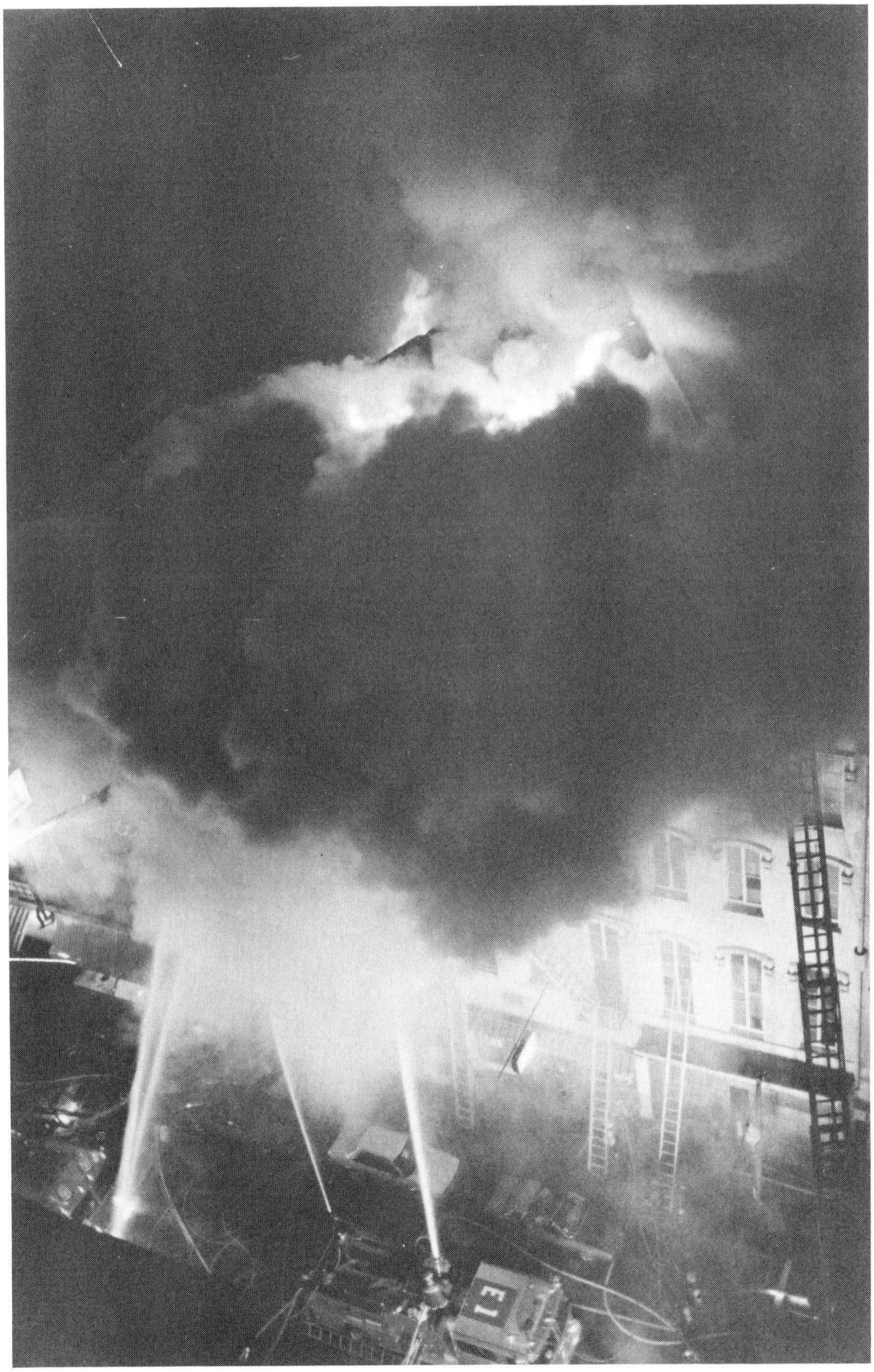

6 Alarm Fire In A Commercial Building
Brian D. Pfister
New Orleans, Louisiana

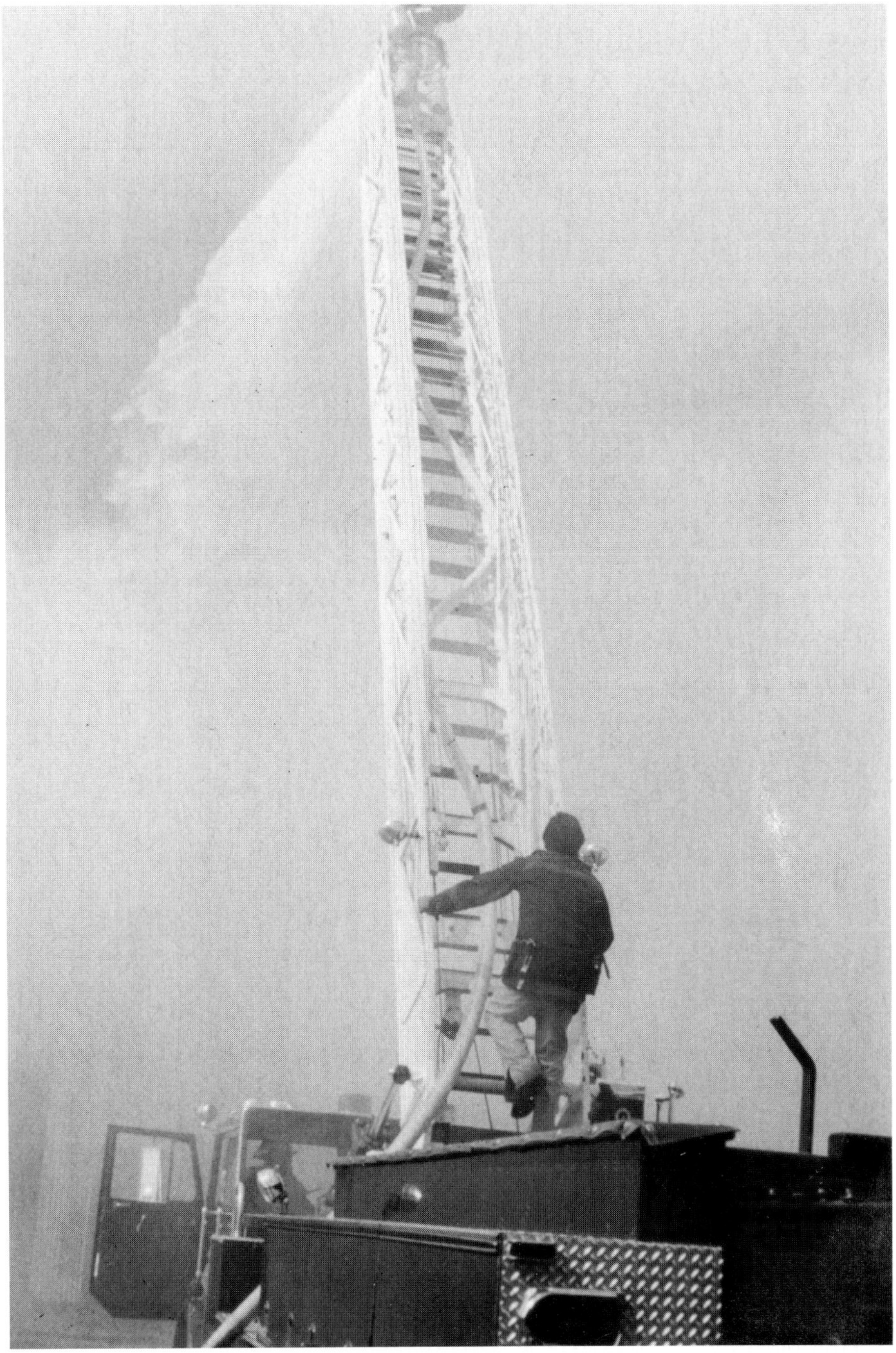

Clark Hotel Greater Alarm
Richard Cowan
Stockton, California

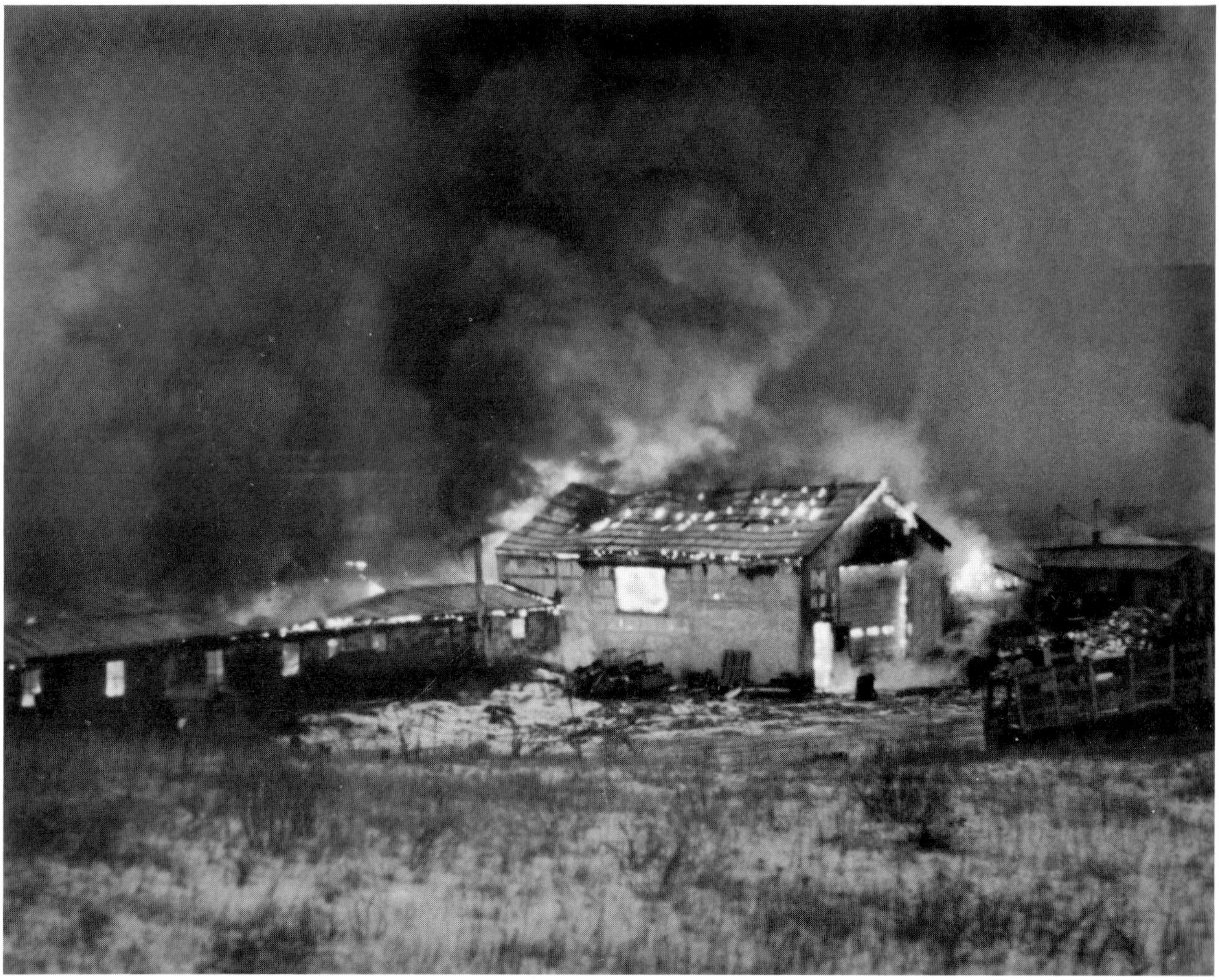

In The Heat of the Night
Chris Slater
Cobleskill, New York

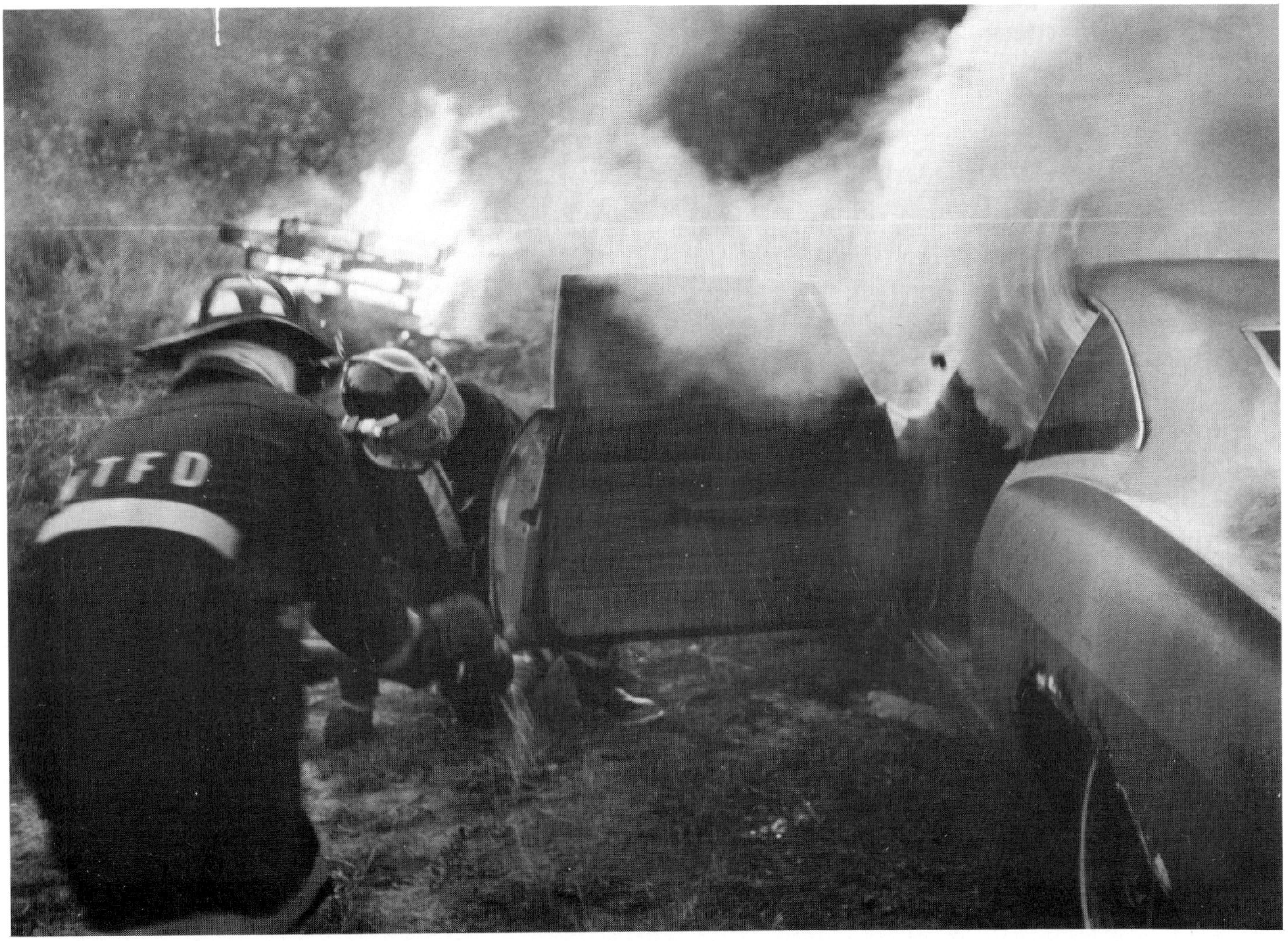

Untitled
Homer Walters
Wooster, Ohio

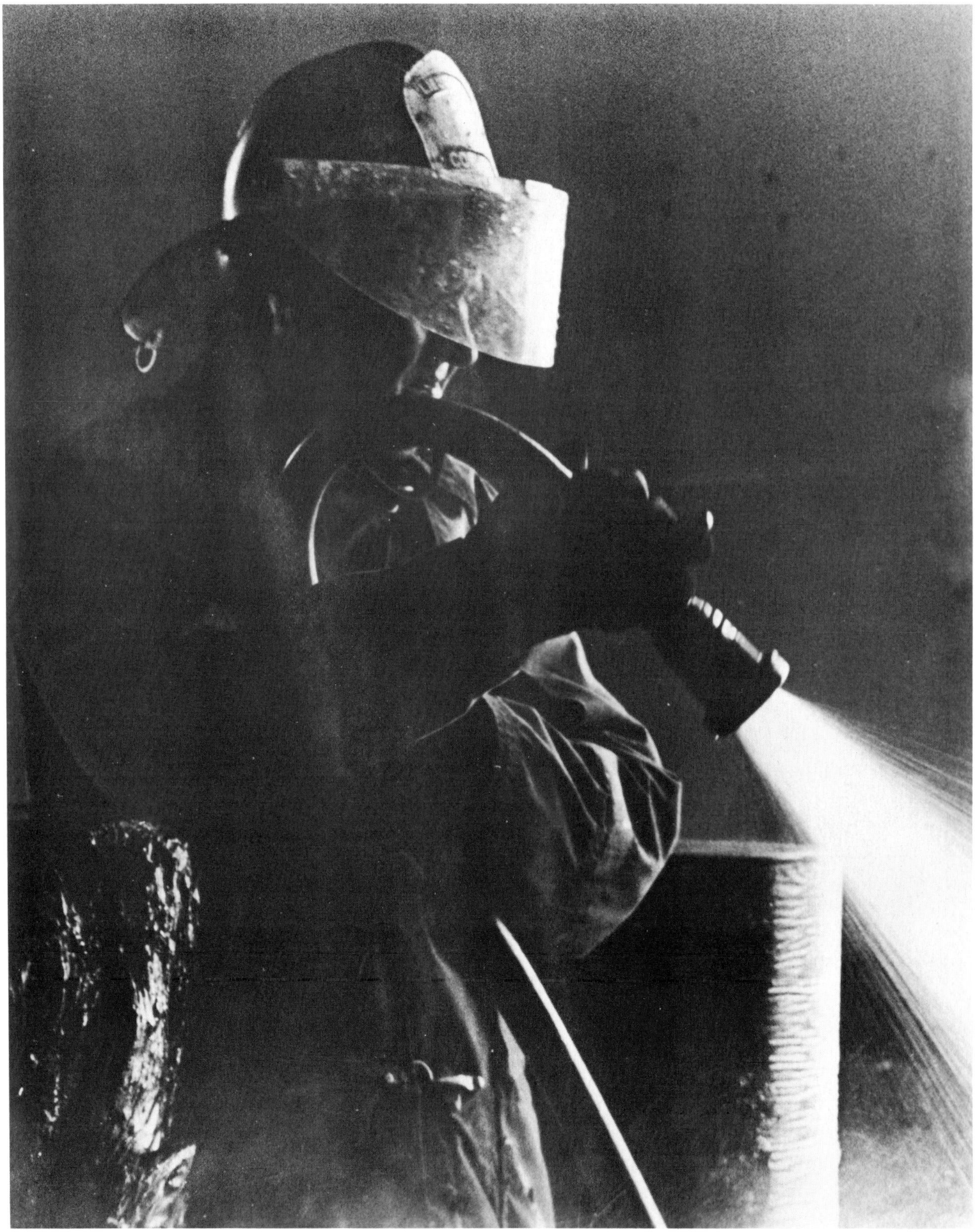

Finishing Touch
George L. King, Jr.
Smyrna, California

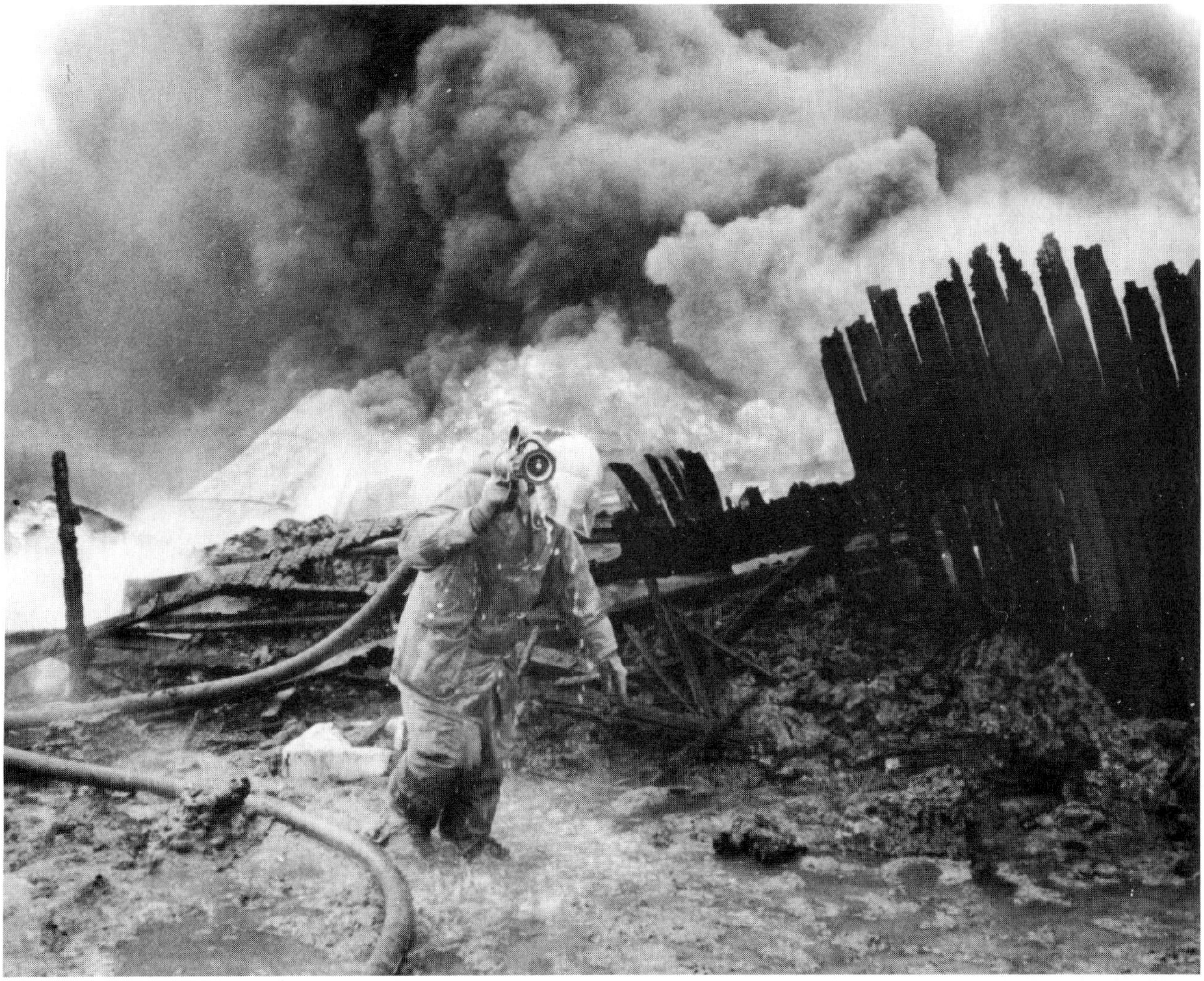

David Erleck, Firefighter
Farmersville, Ohio

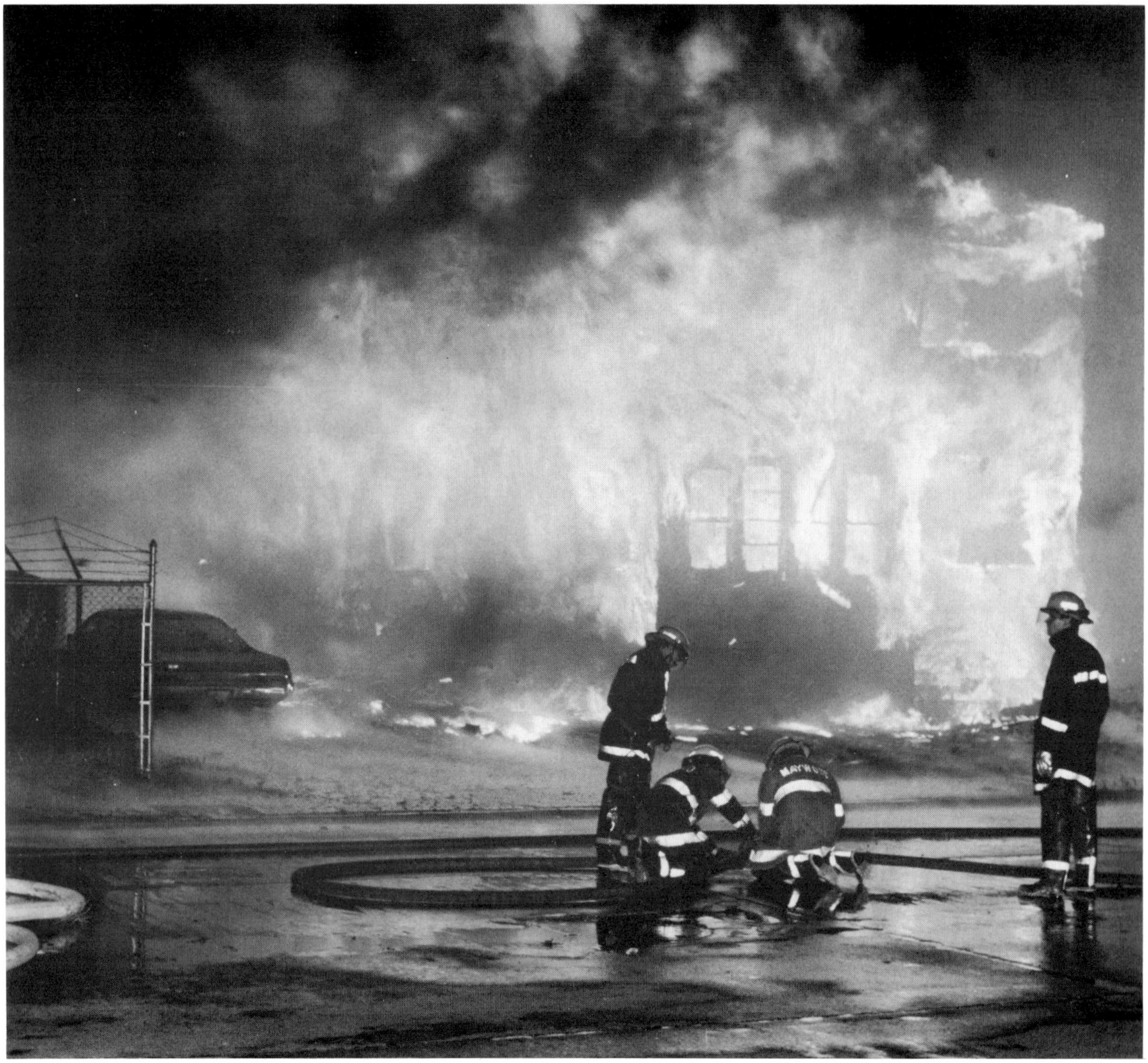

Surround & Drown
Paul Gardner, Jr.
Berwyn, Illinois

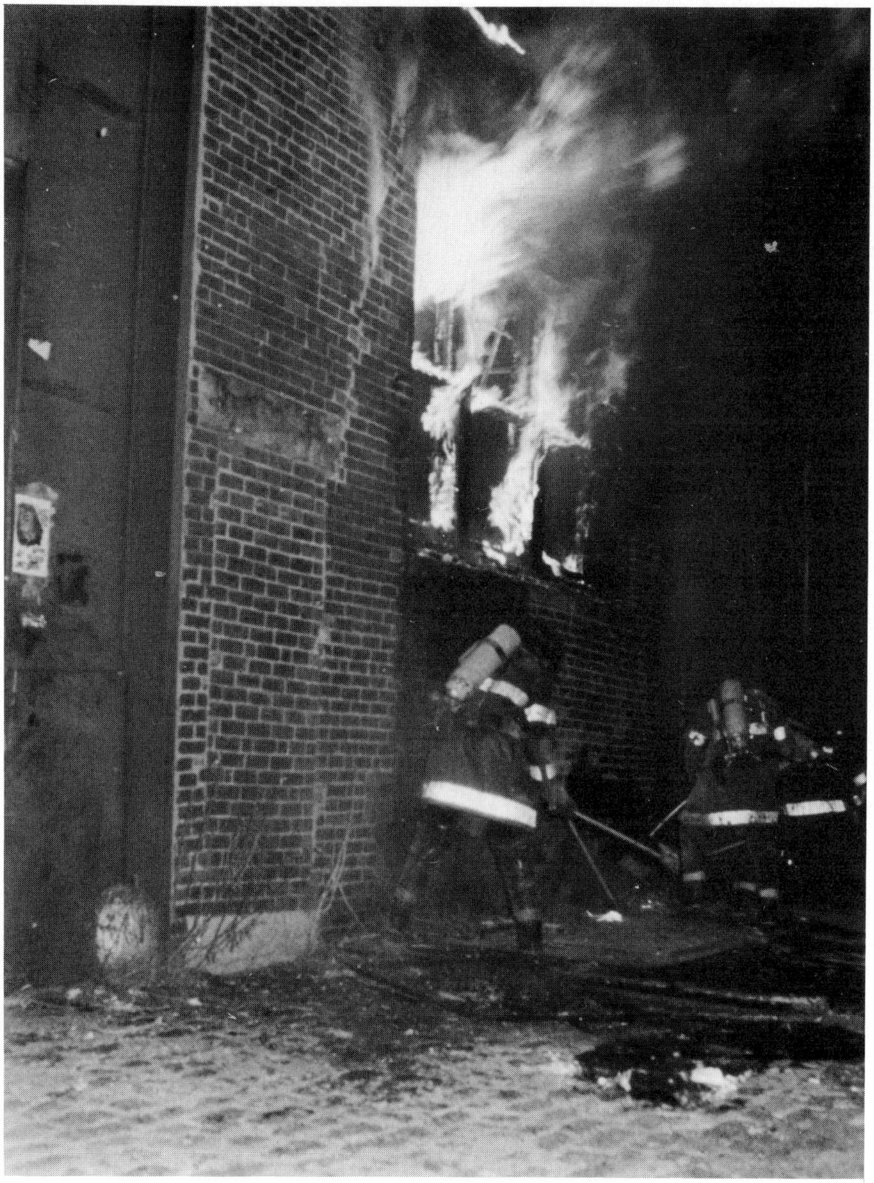

Opening Up
Joseph Slattery
Natick, Massachusetts

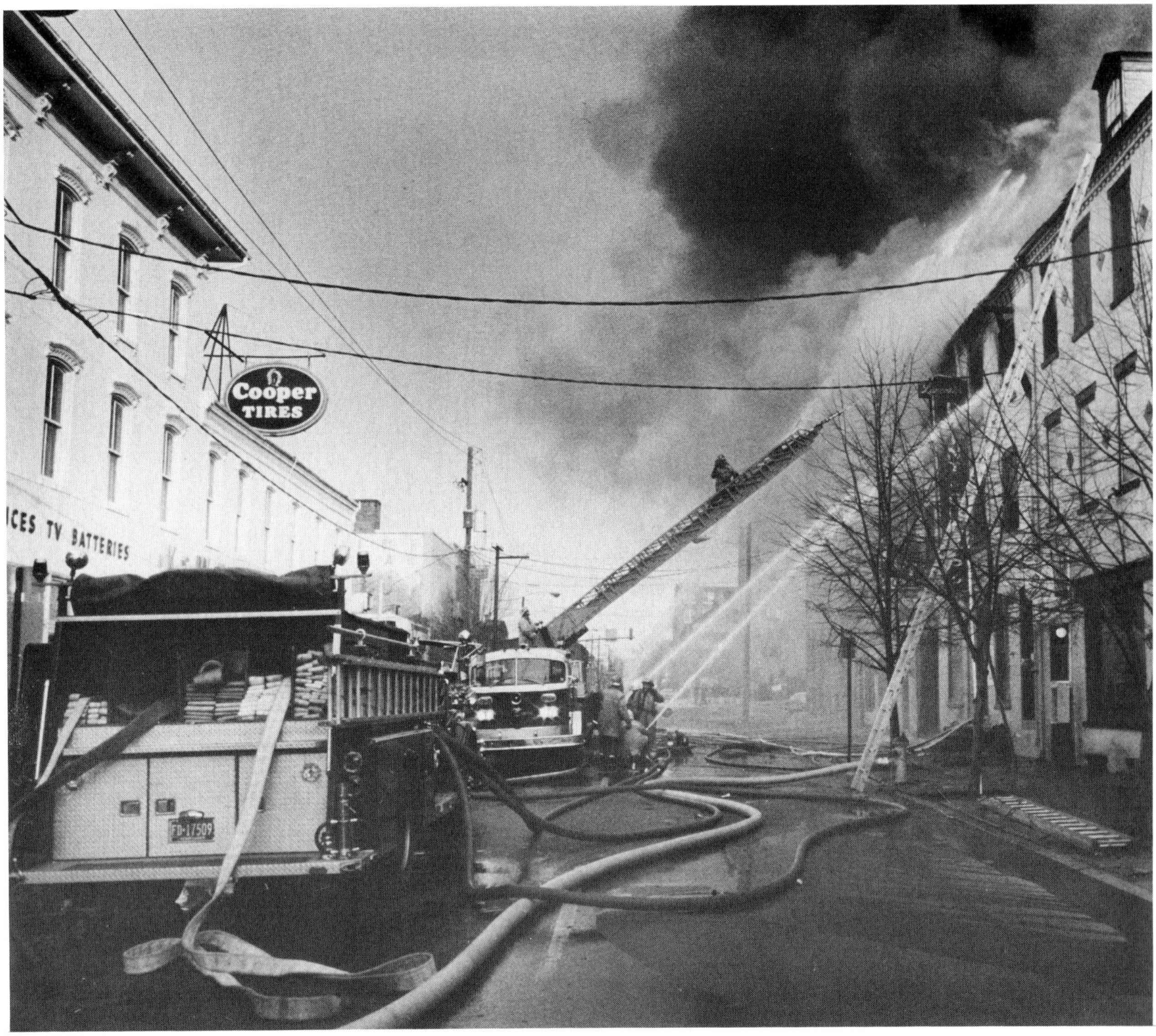

Steven A. Schnoke
Schuylkill Haven, Pennsylvania

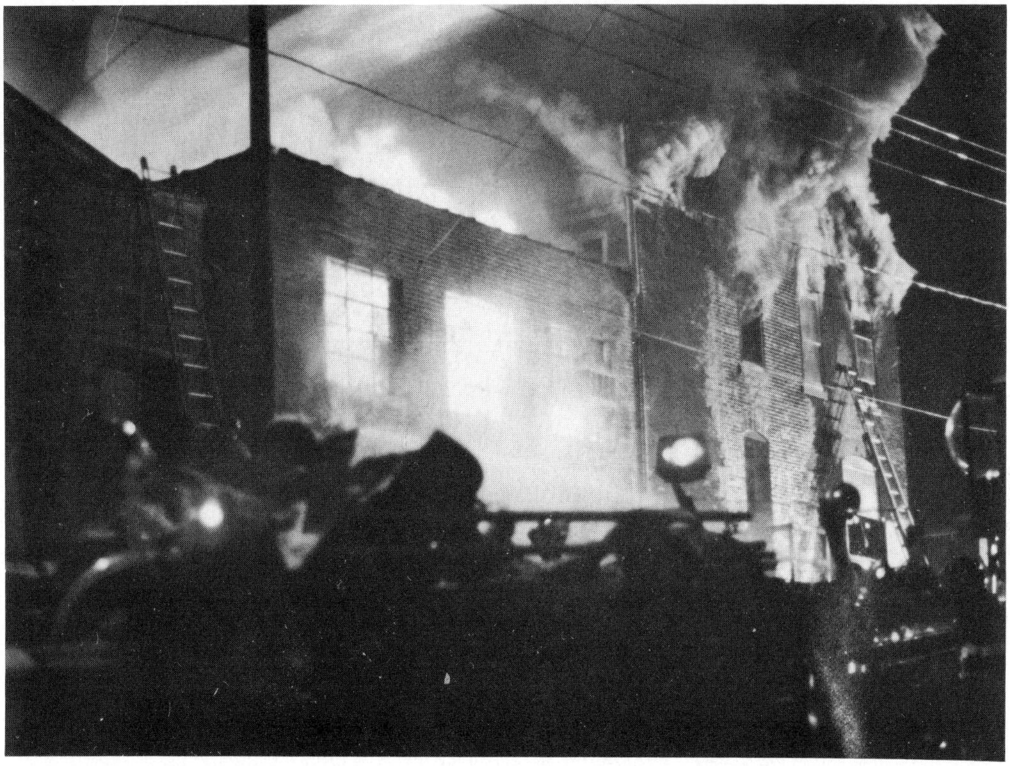

Streams of Suppression
Blaine T. Shahan
Lancaster, Pennsylvania

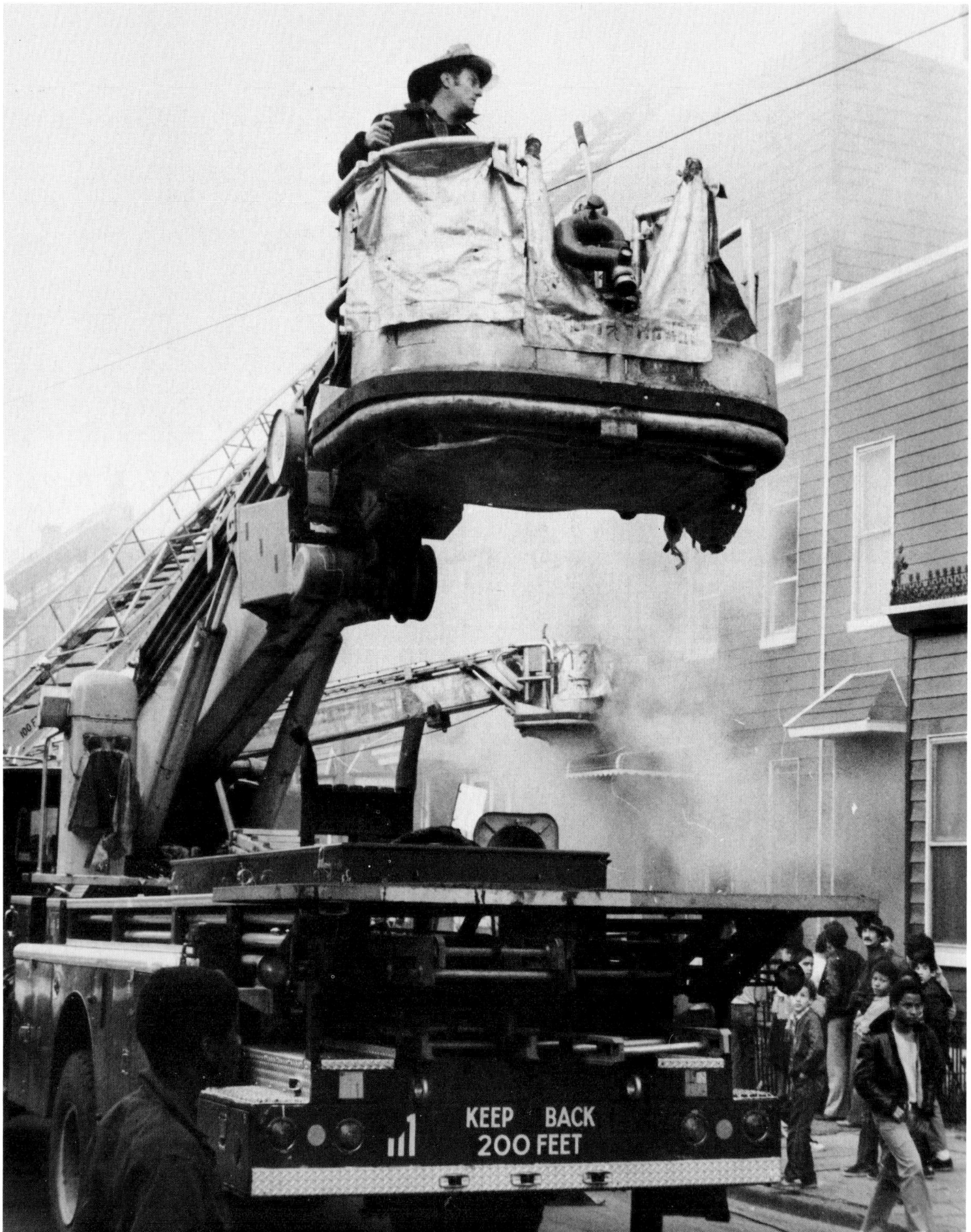

Chaffer Goes To Adjacent
Roof . . . Everyone Work!
James V. Sanquedolce
Mystic, Connecticut

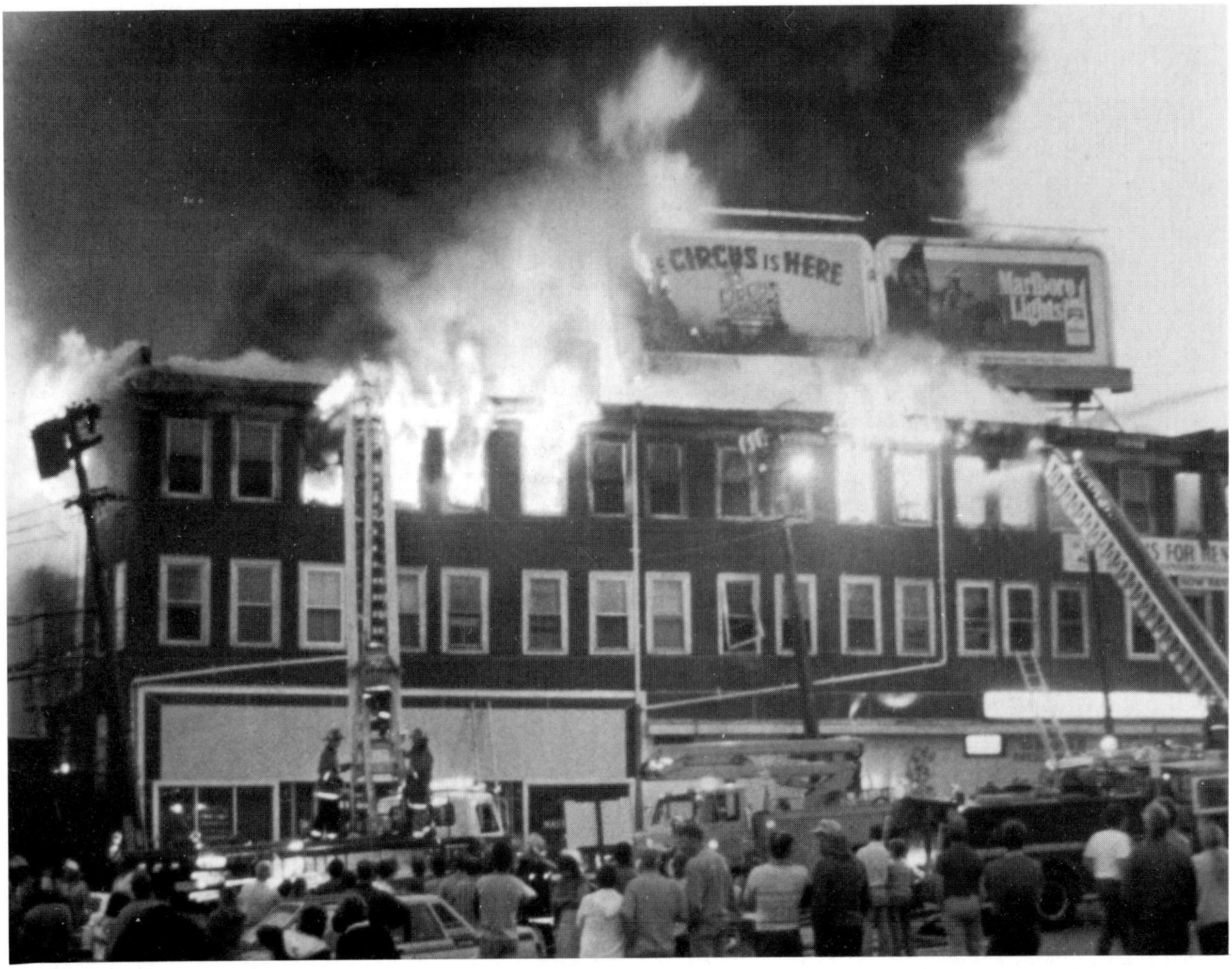

Heavy Fire
Robert D. Battis
Beverly, Massachusetts

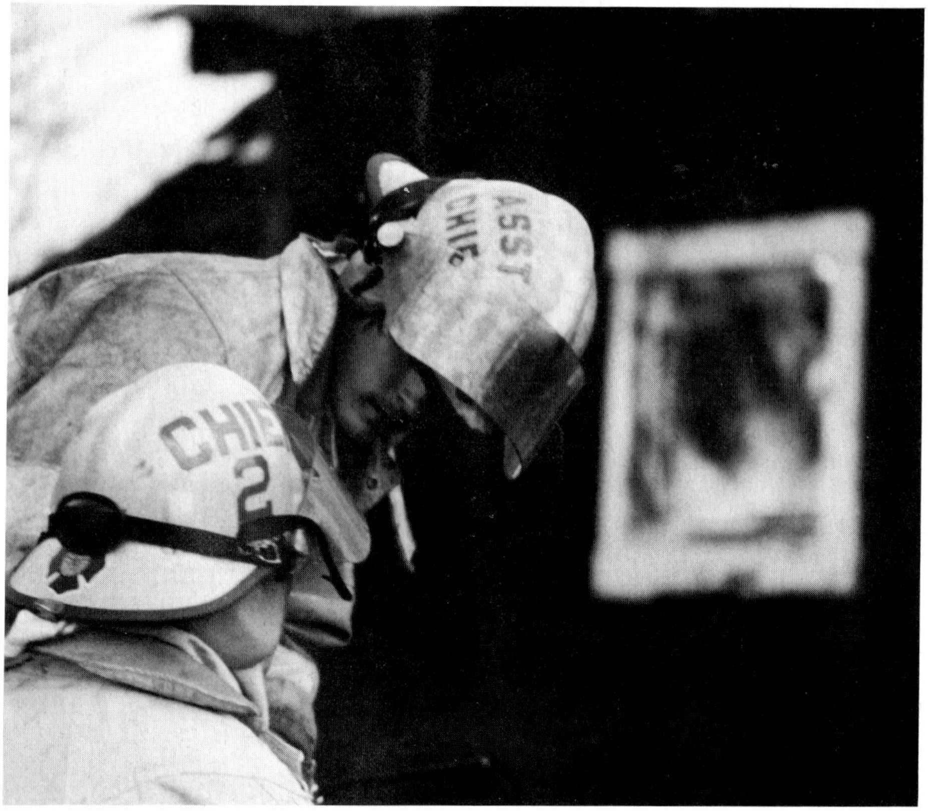

Conference
Joseph Maringo
Pittsburgh, Pennsylvania

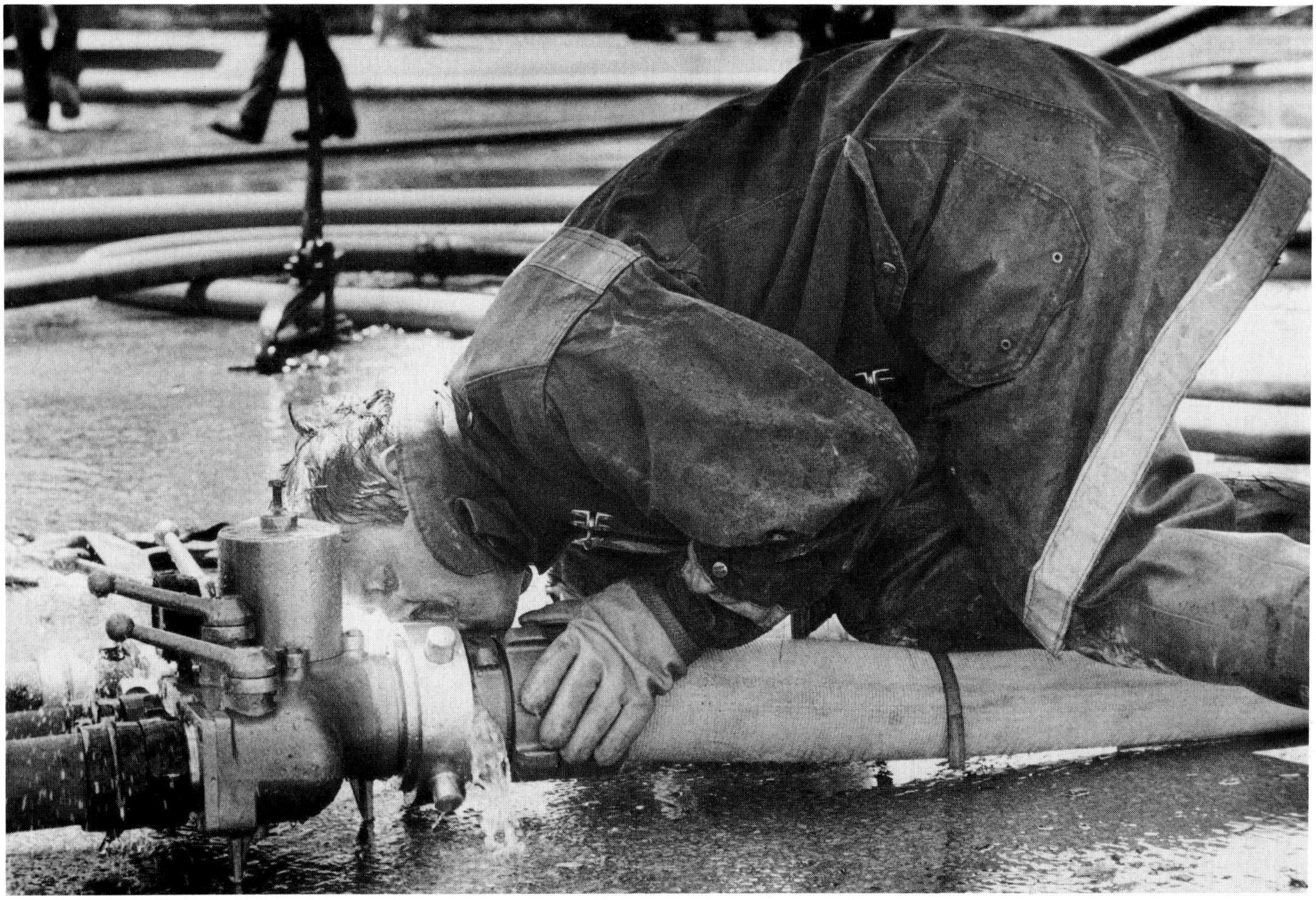

Taking a Gulp
David Perry
Lexington Herald-Leader
Lexington, Kentucky

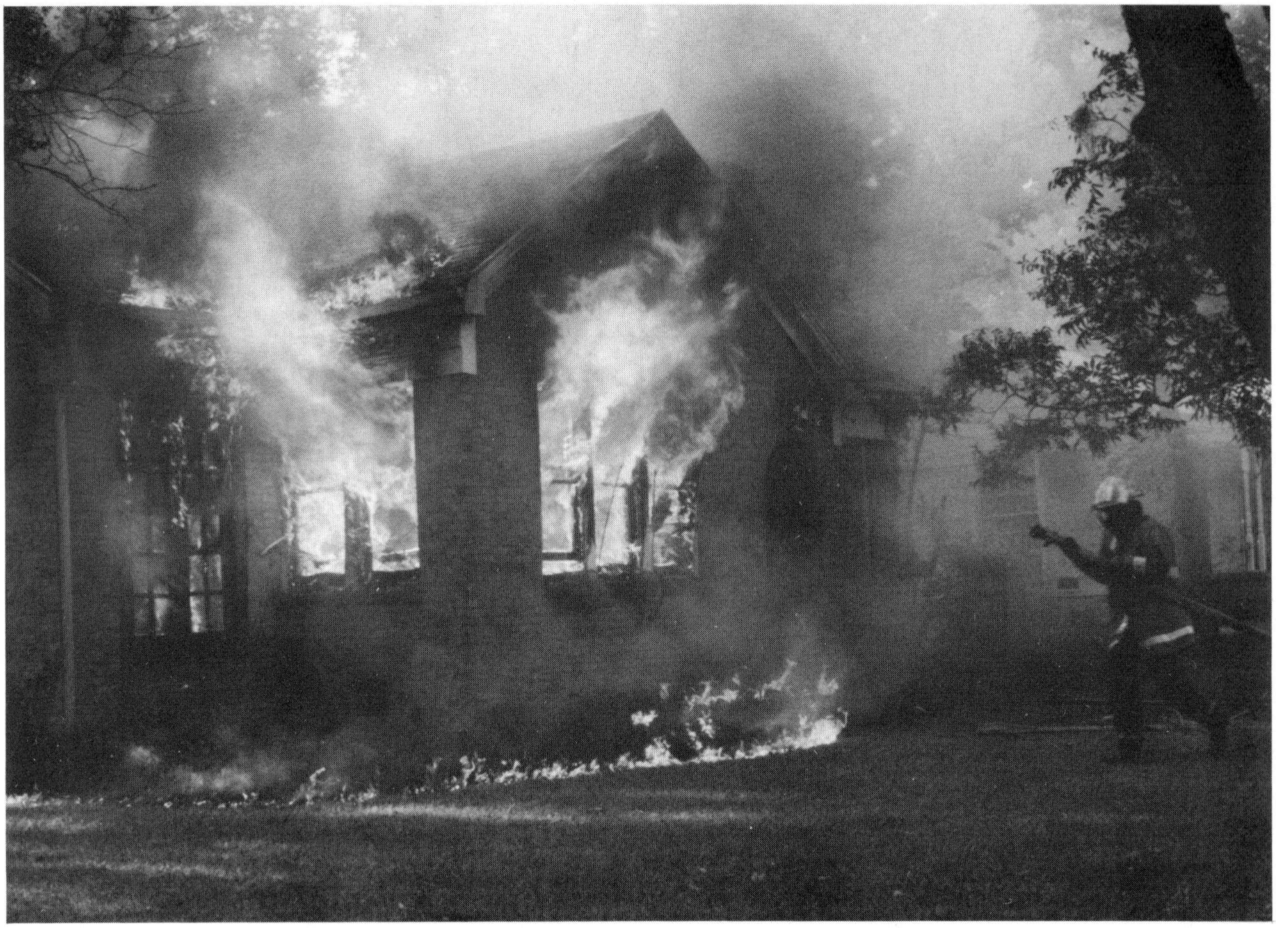

James M. Hulen
Sulphur Springs, Texas

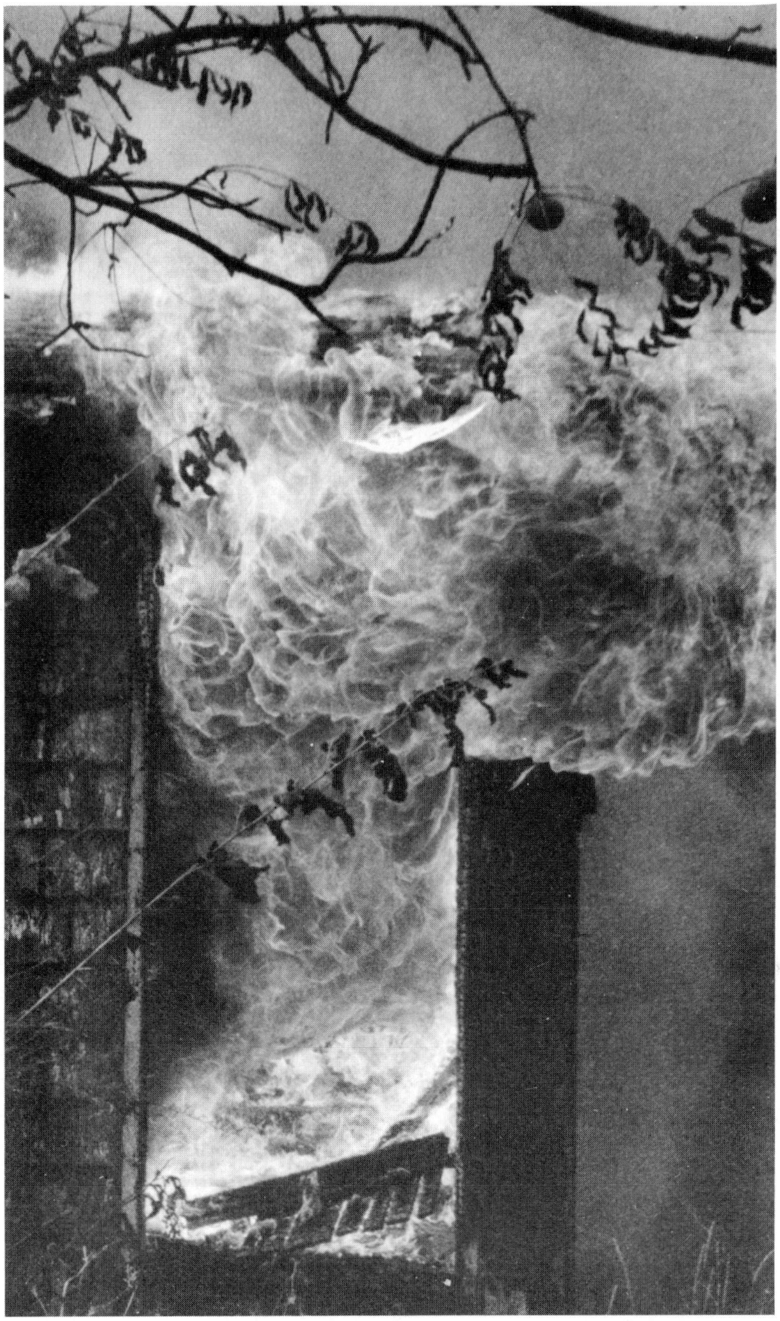

Fire Showing
David E. Rucker
Brandywine, Maryland

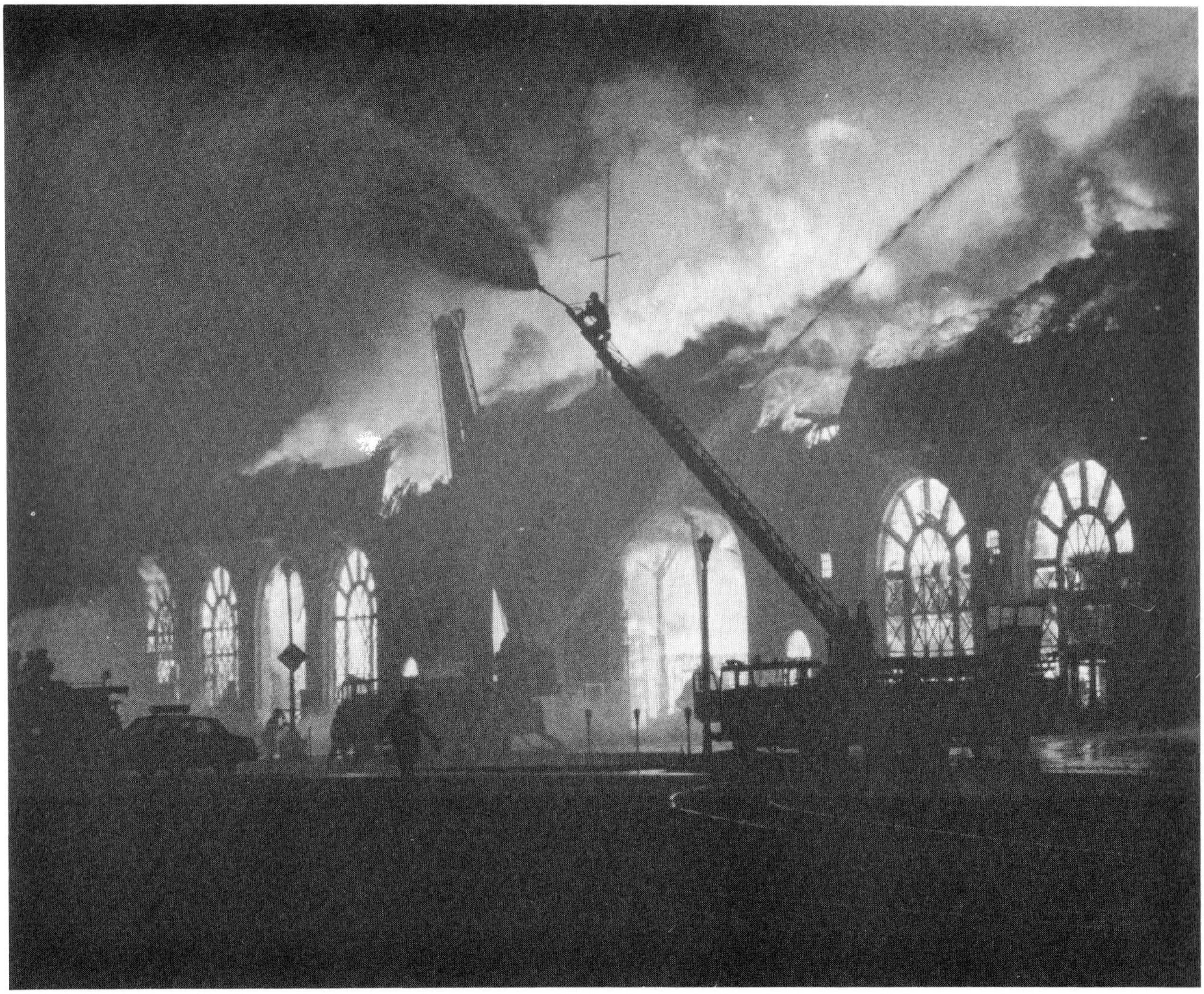

Milton Yuen
San Francisco, California

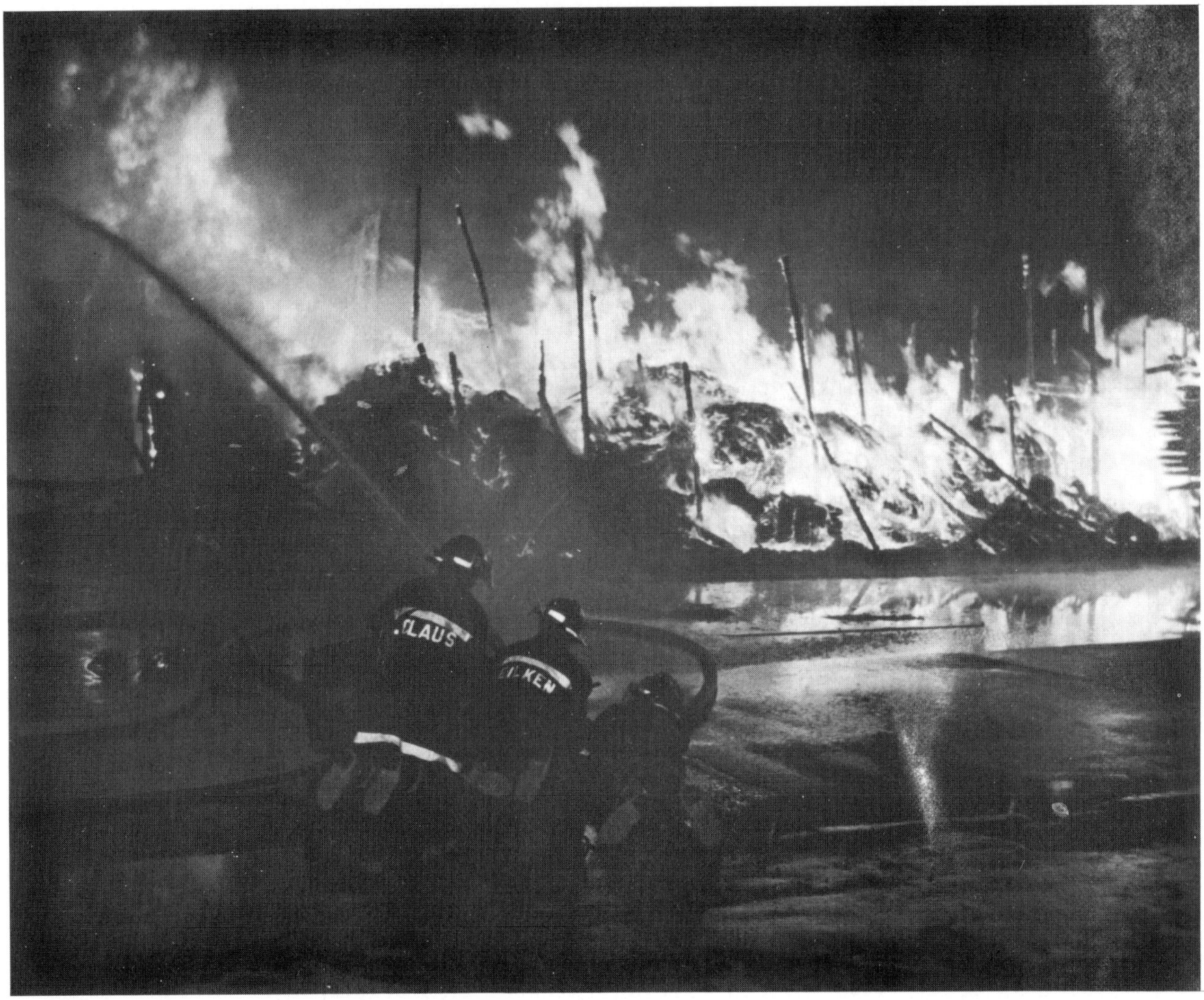

David Berger
Northbrook, Illinois

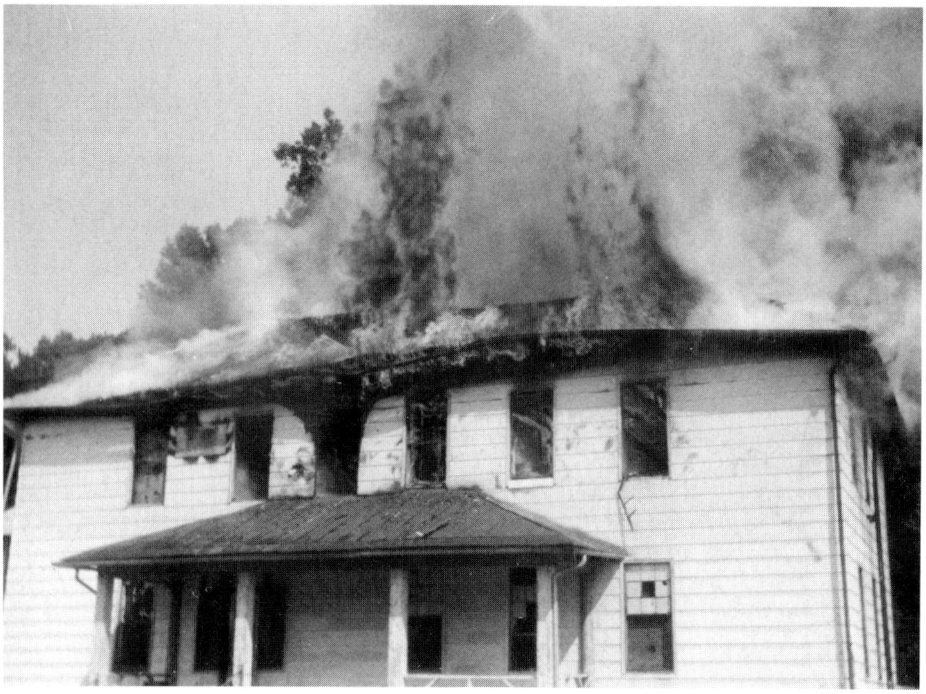

Clare Railroad Yard
Jeff Root
Cincinnati, Ohio

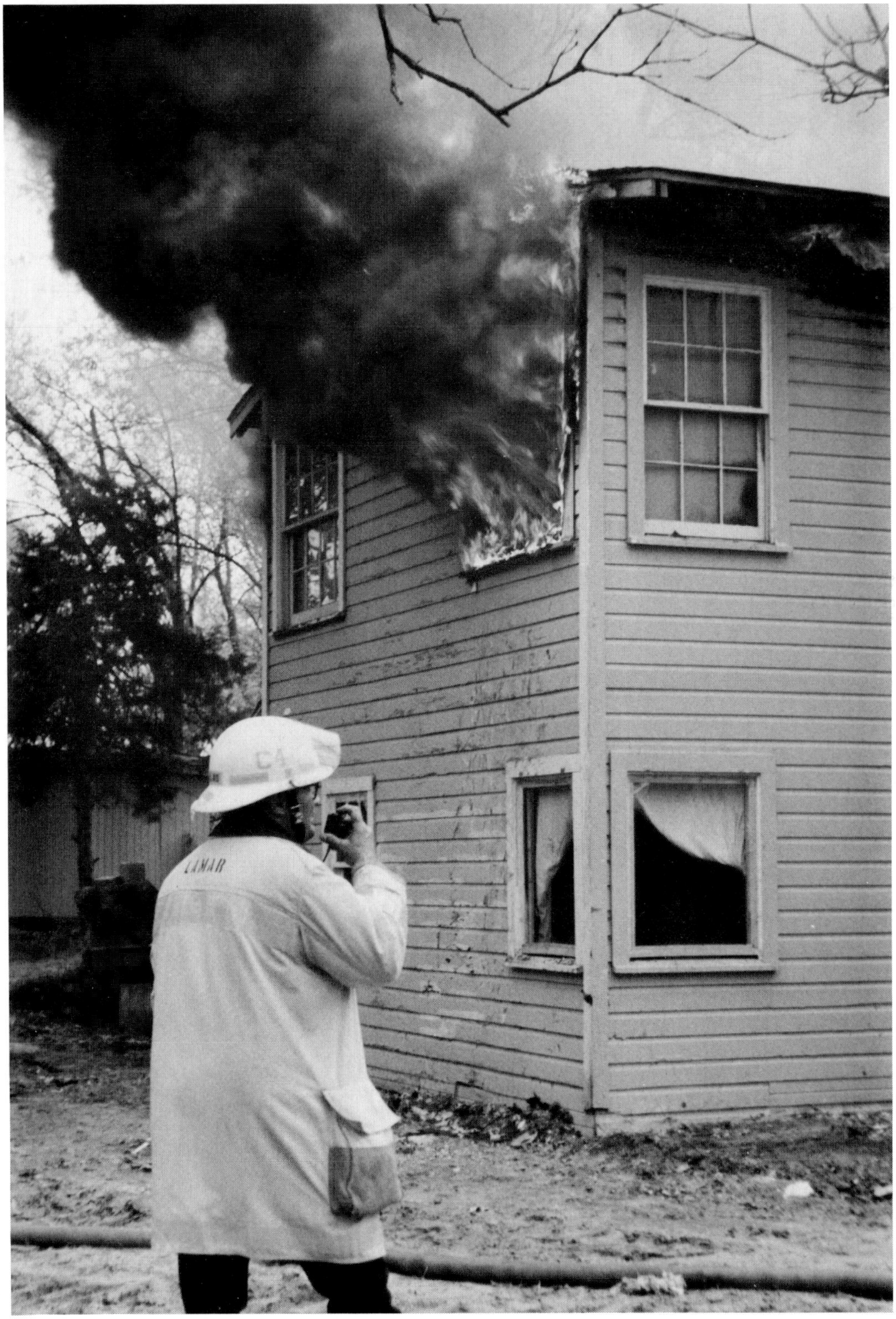

Bring The Water
Karen Deering
Mesquite, Texas

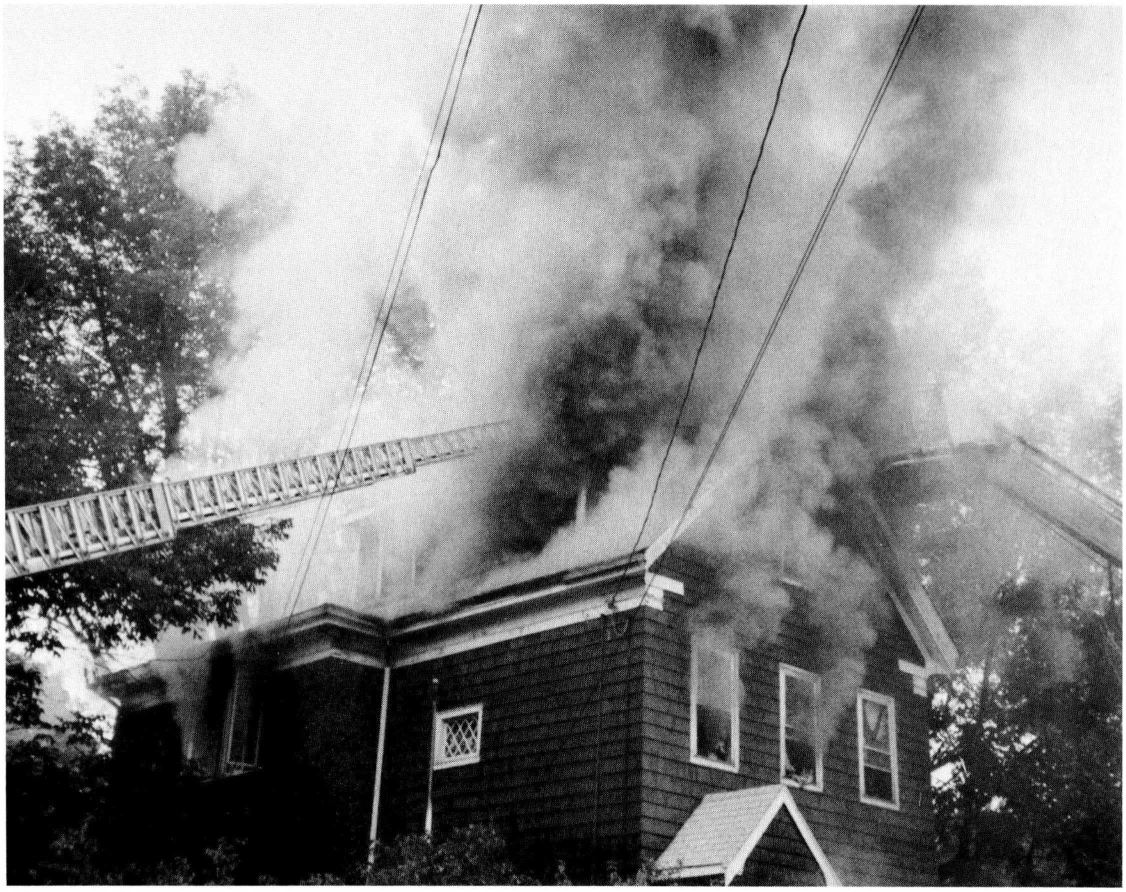

John Antonelli
Somerville, Massachusetts

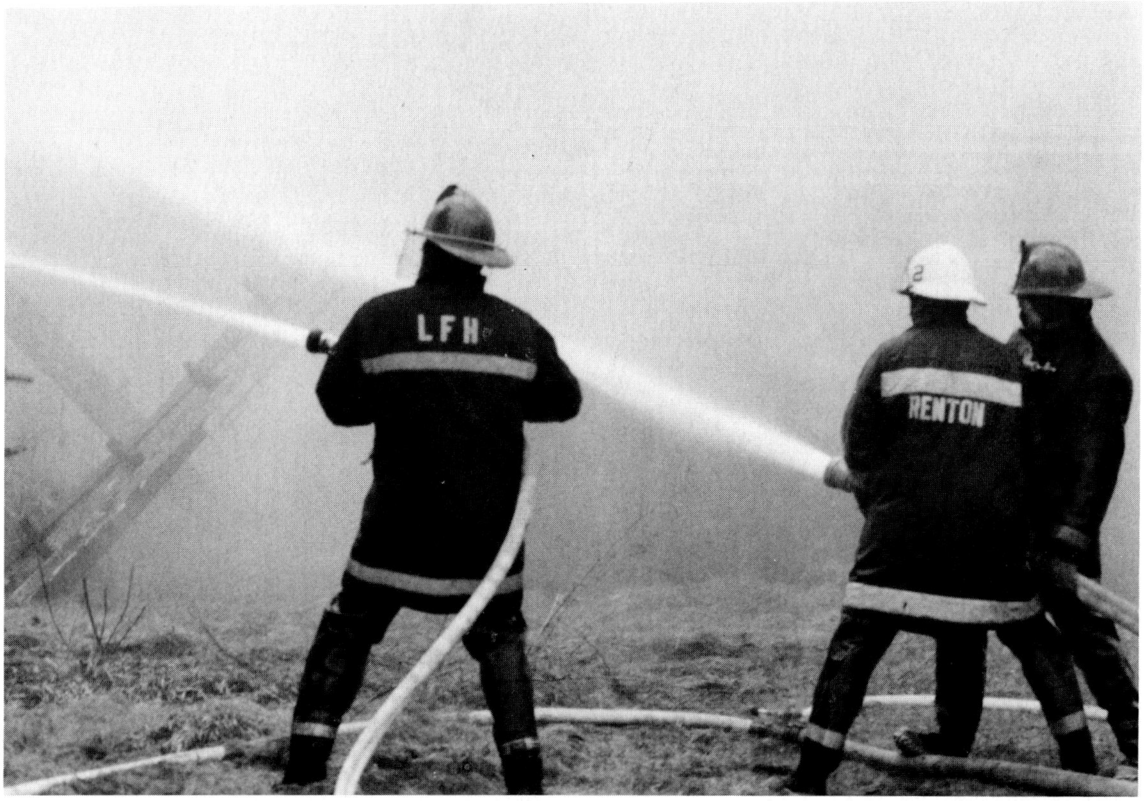

Teamwork
Barb Maringo
Pittsburgh, Pennsylvania

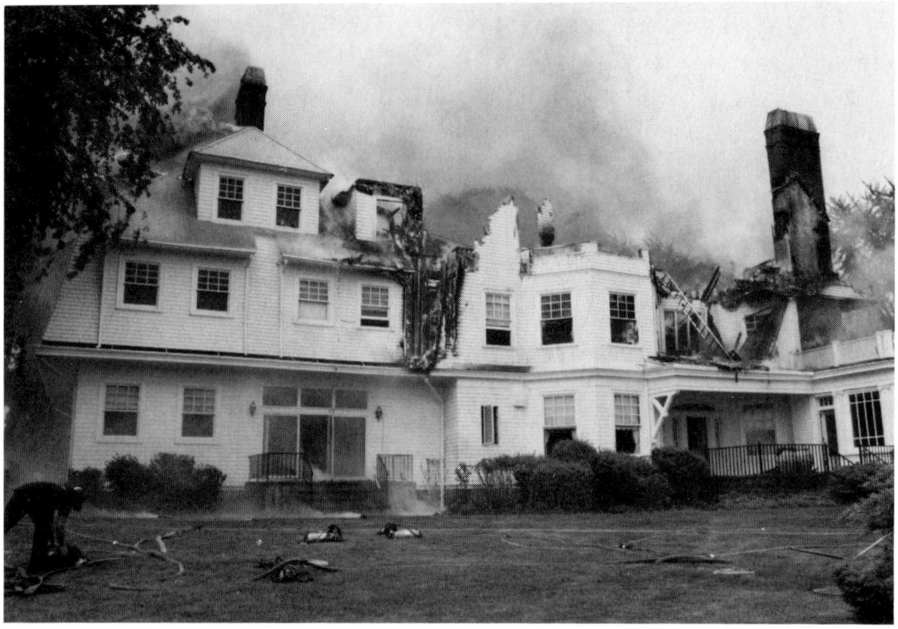

John P. Maguire
New Rochelle, New York

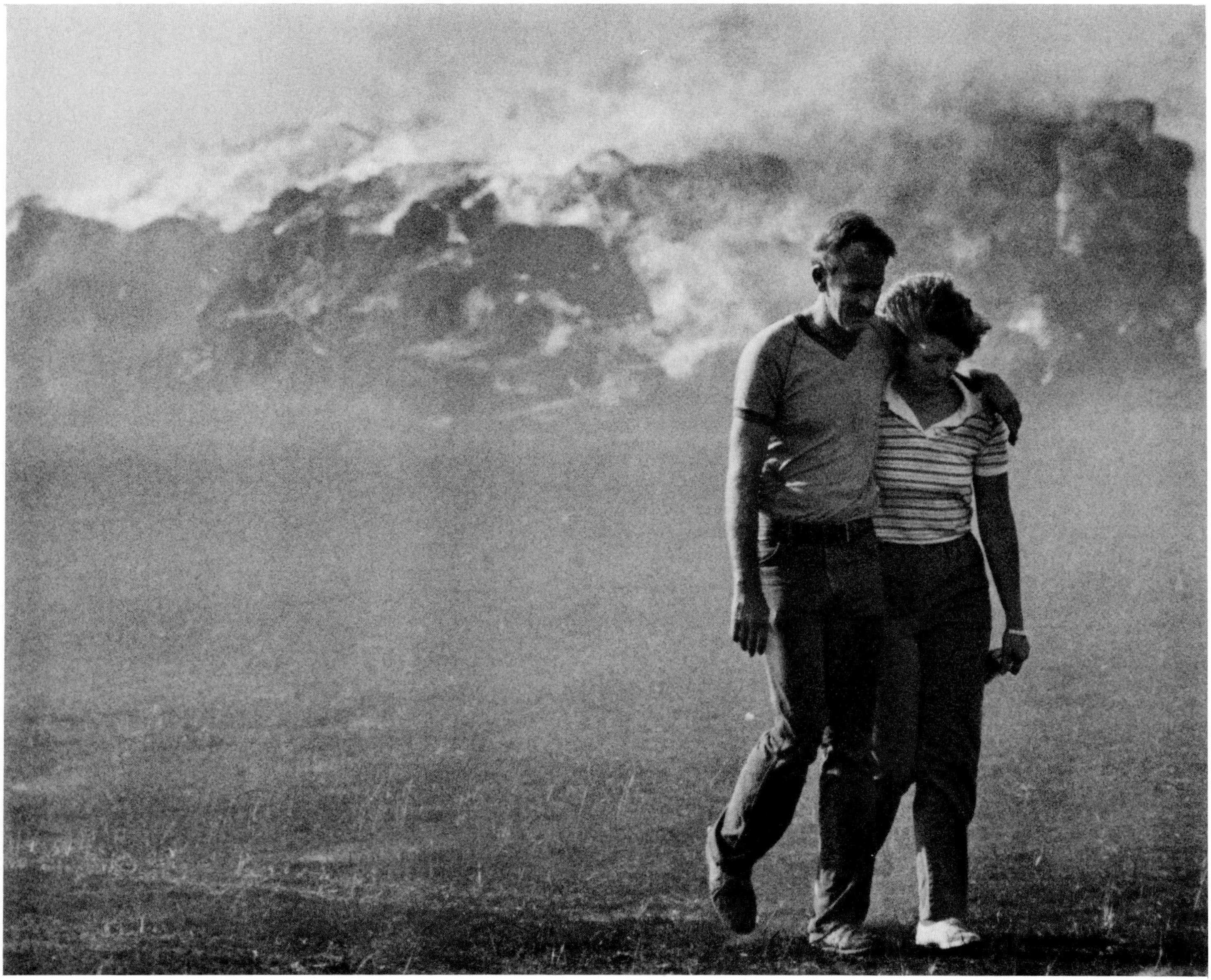

Losing It All
Larry R. Fleming
Wichita Eagle-Beacon
Wichita, Kansas

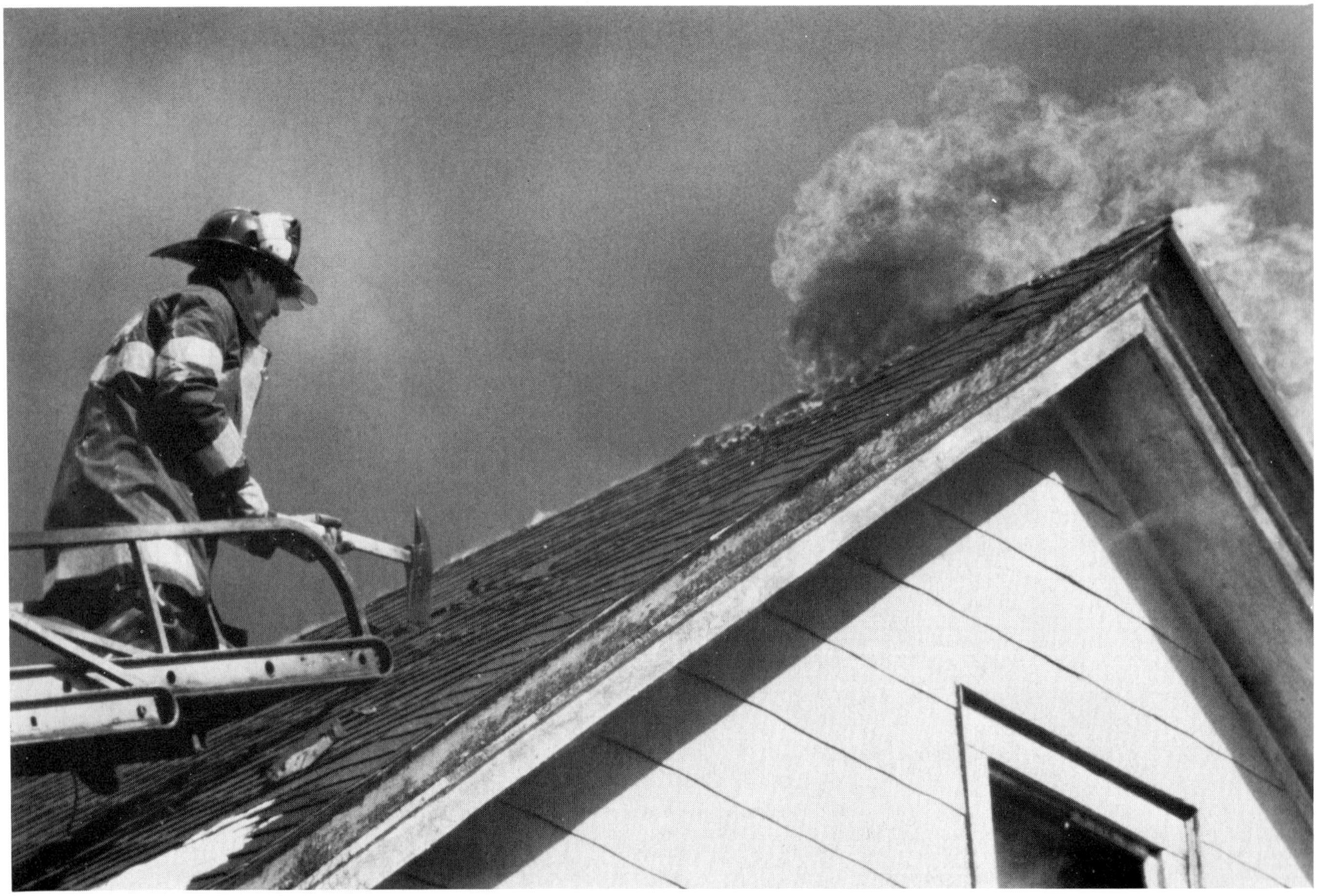

Truckie Proud of his Work
Paul La Rochelle, Jr.
Southbridge, Massachusetts

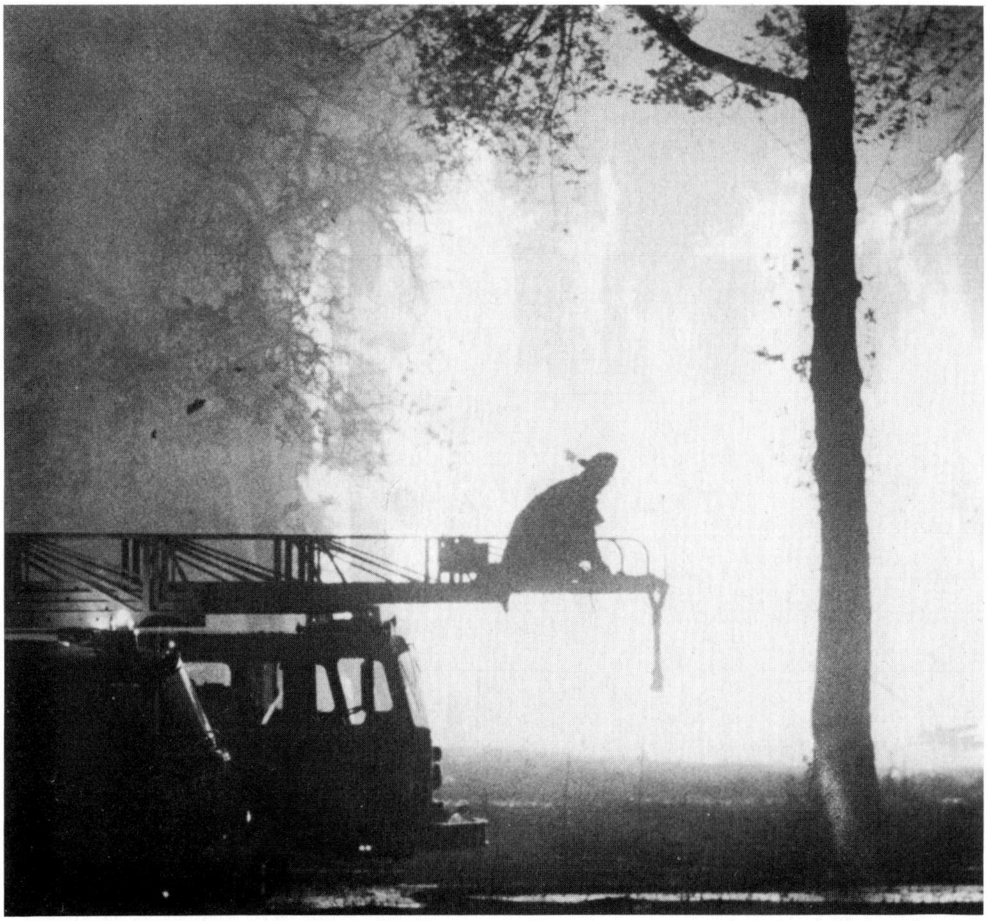

Alone & Unafraid
Joseph E. Provost
Dearborn, Michigan

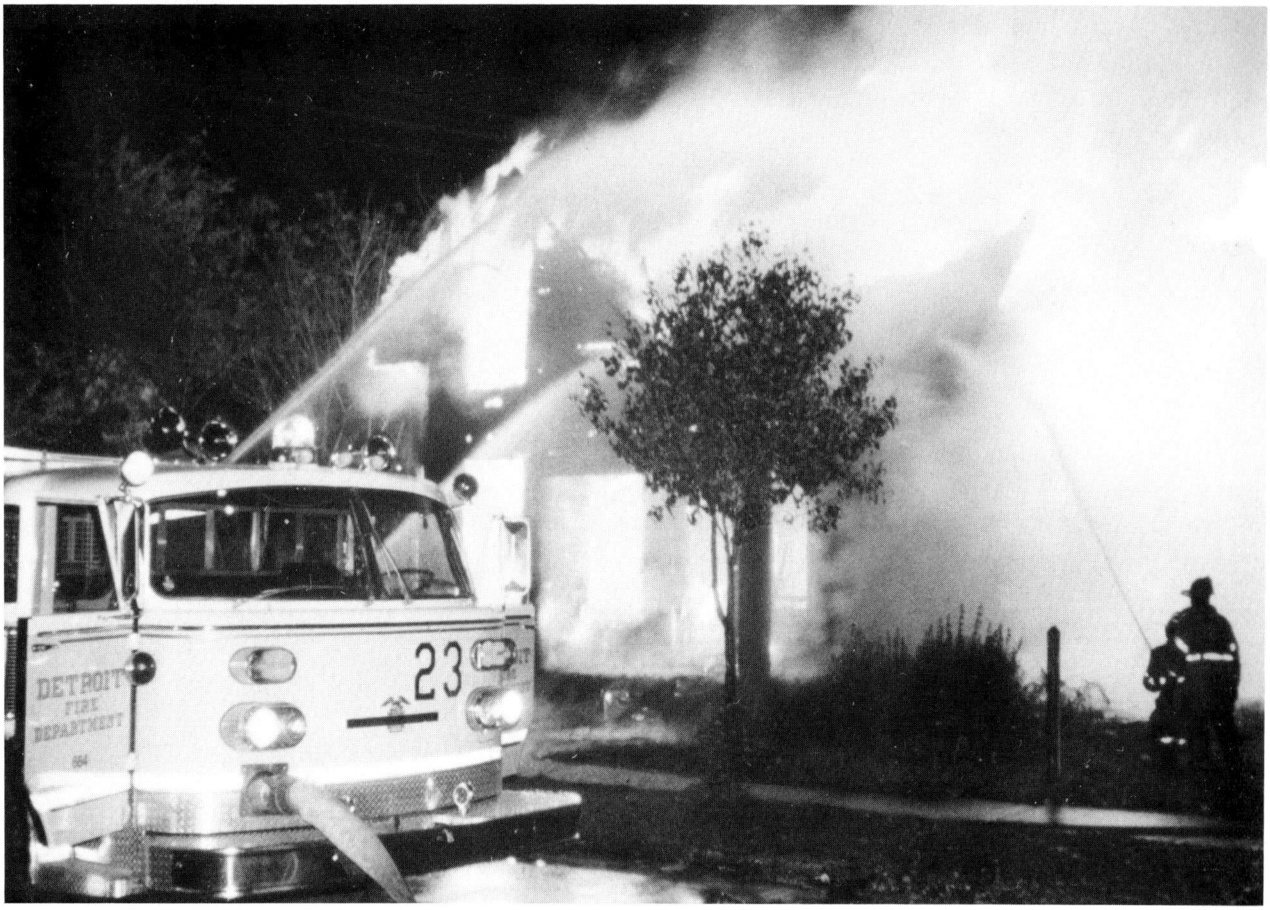

Devil's Night '84
Robert Paleczny
Elizabeth, New Jersey

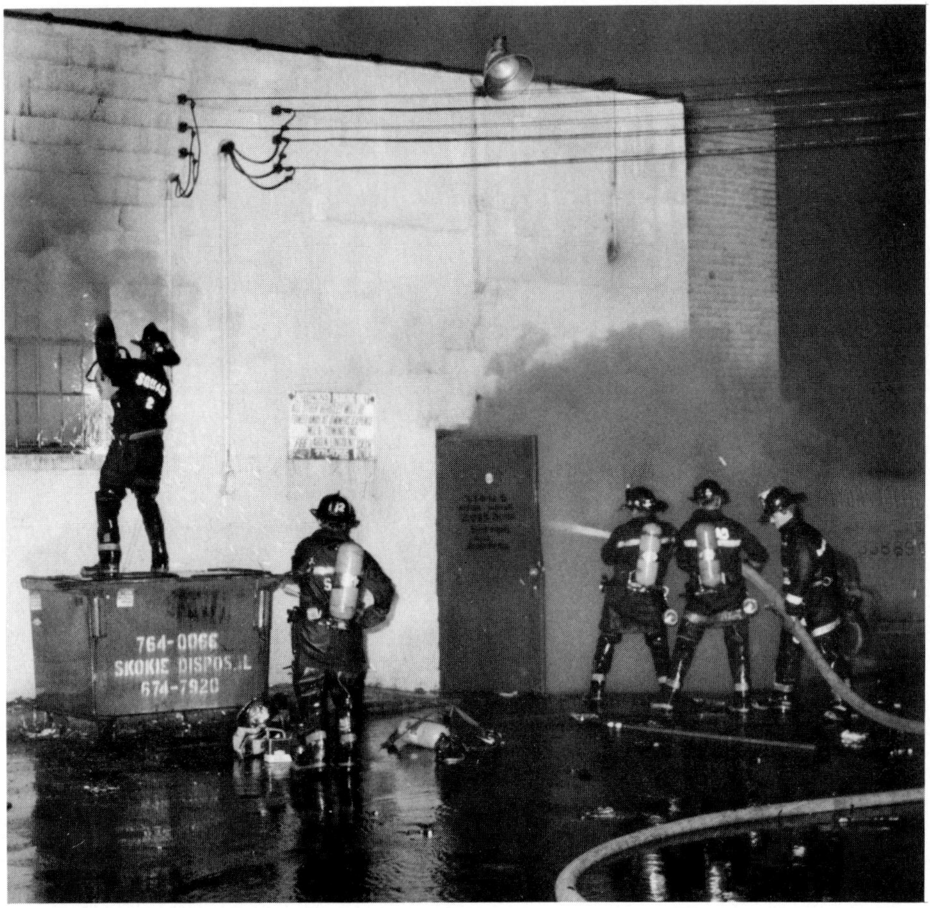

Chicago Squadies at Work
Gordon J. Nord, Jr.
Lyons, Illinois

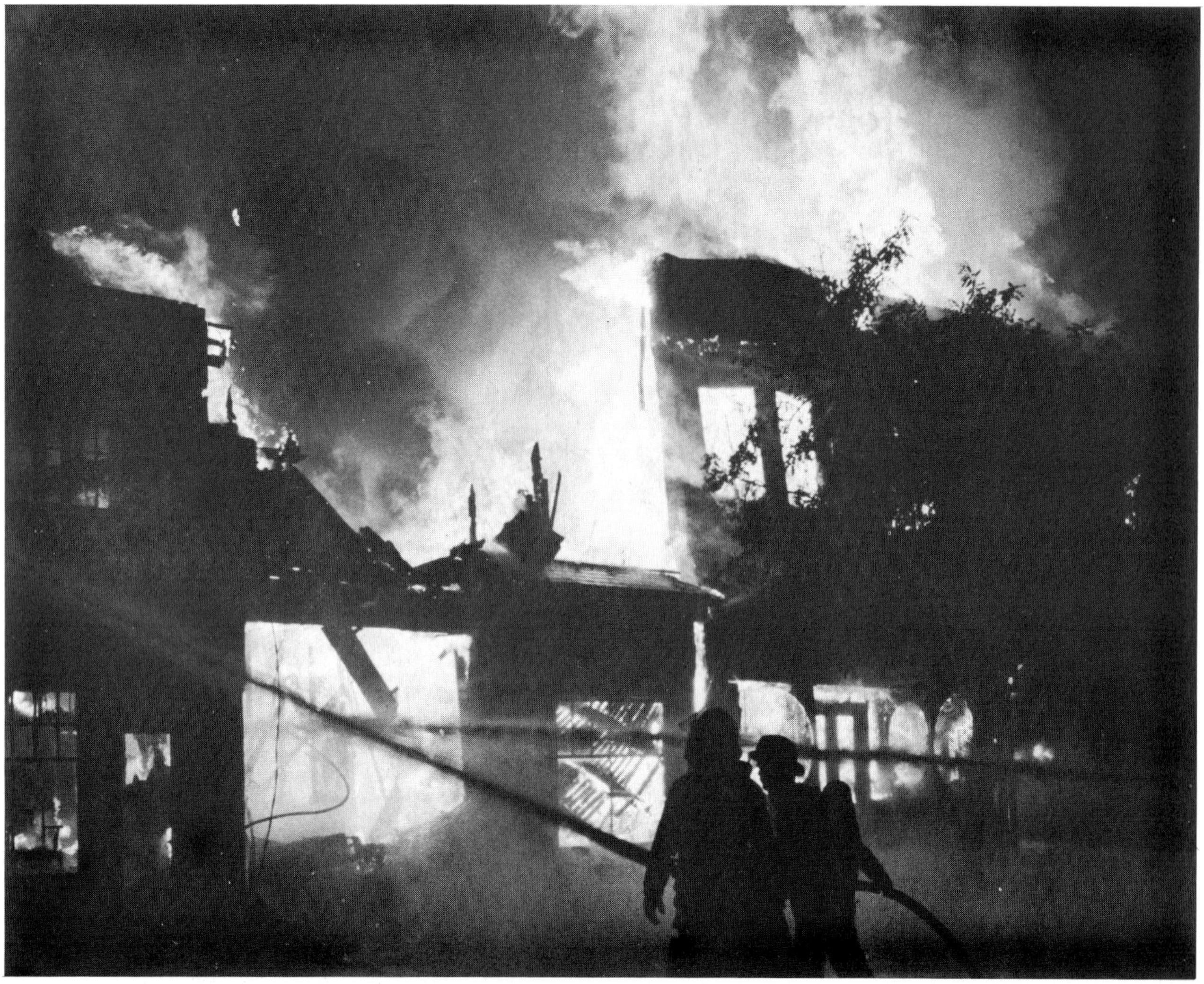

Million Dollar Fire
Scott Myers
Mill Valley, California

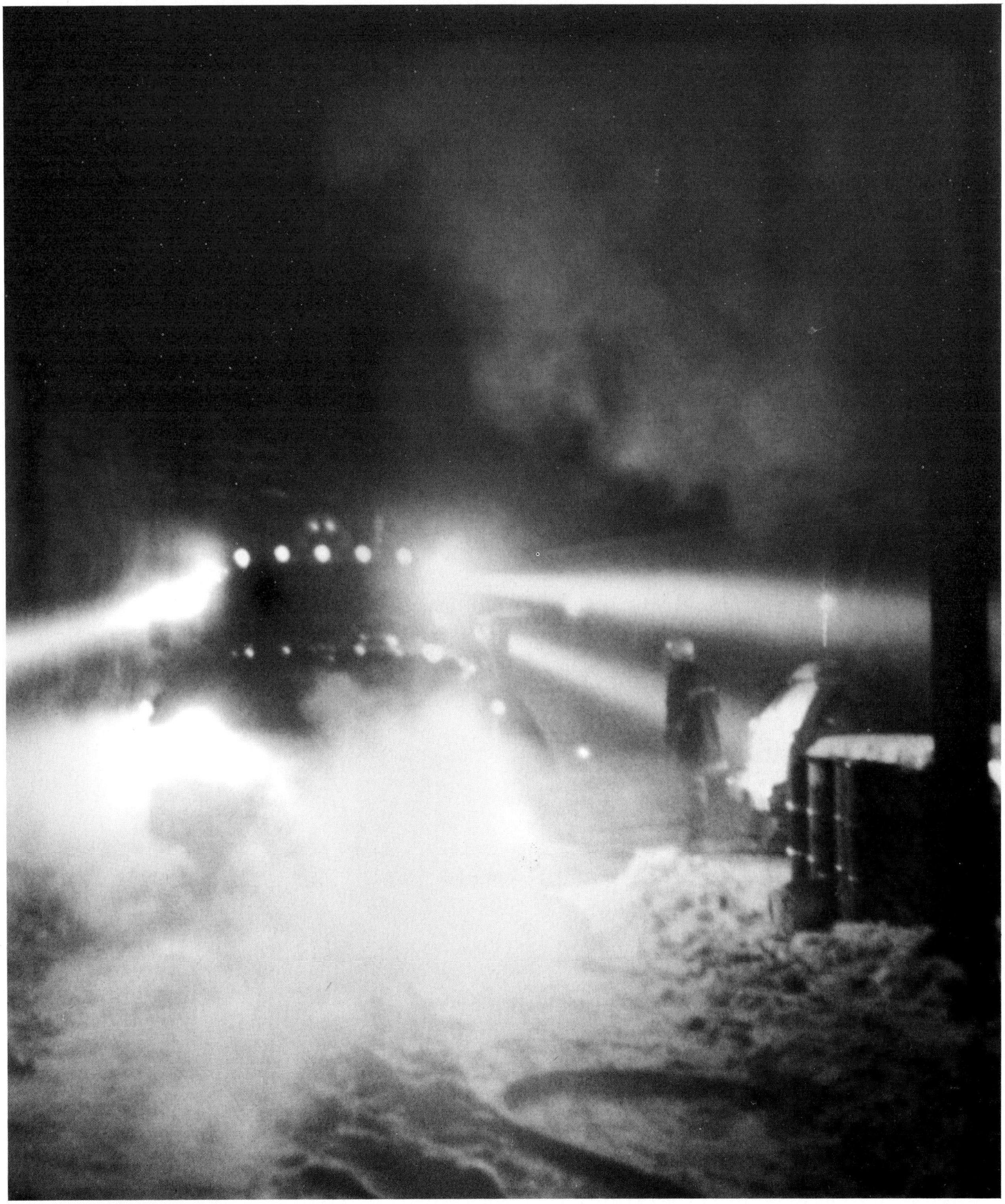

Fire At 15 Degrees
Scott Gregory
Lewistown, Montana

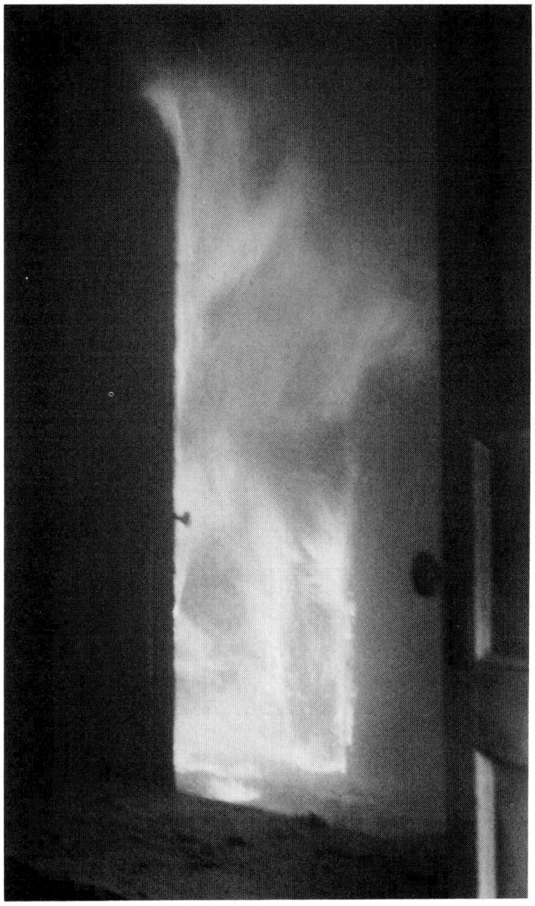

The Devil's Doorway
Jim Carrier
Saukville, Wisconsin

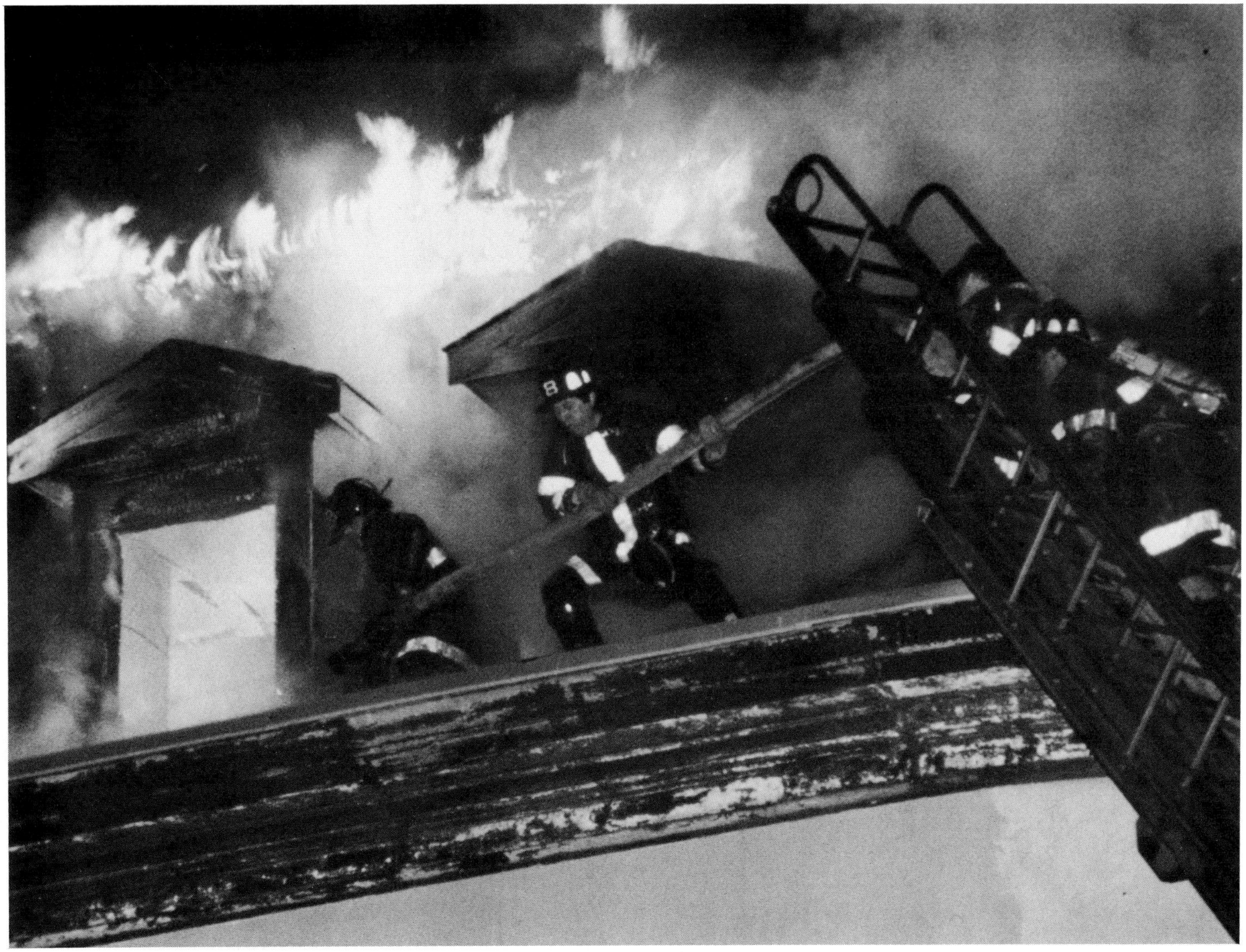

Stretching In
Paul LaRochelle, Jr.
Southbridge, Massachusetts

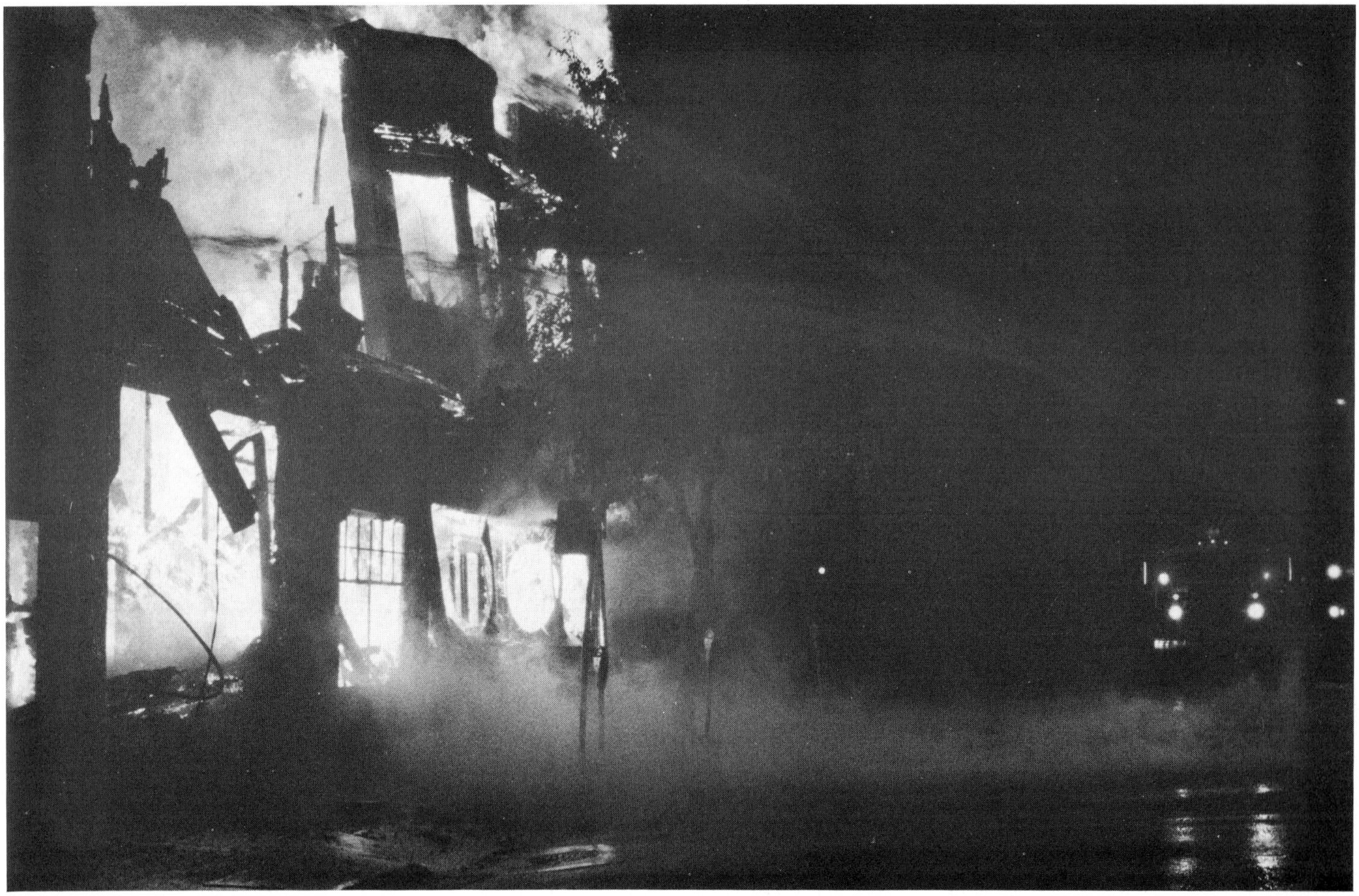

Stoughton Box 3-3
Robert Myers
Brockton, Massachusetts

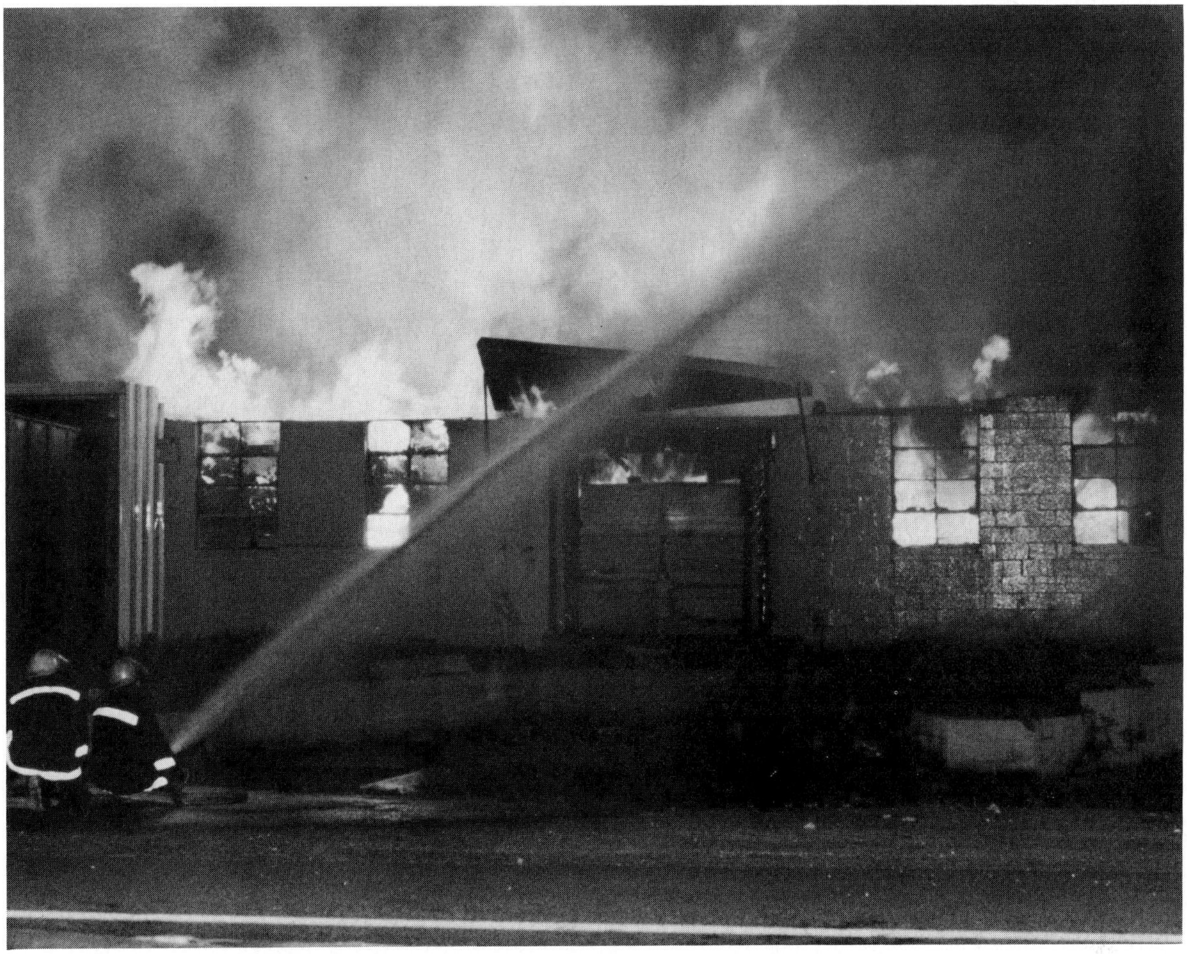

Raging Fire Claims —
$2 Million Dollar Loss
Dreama Floersch
Milan, Tennessee

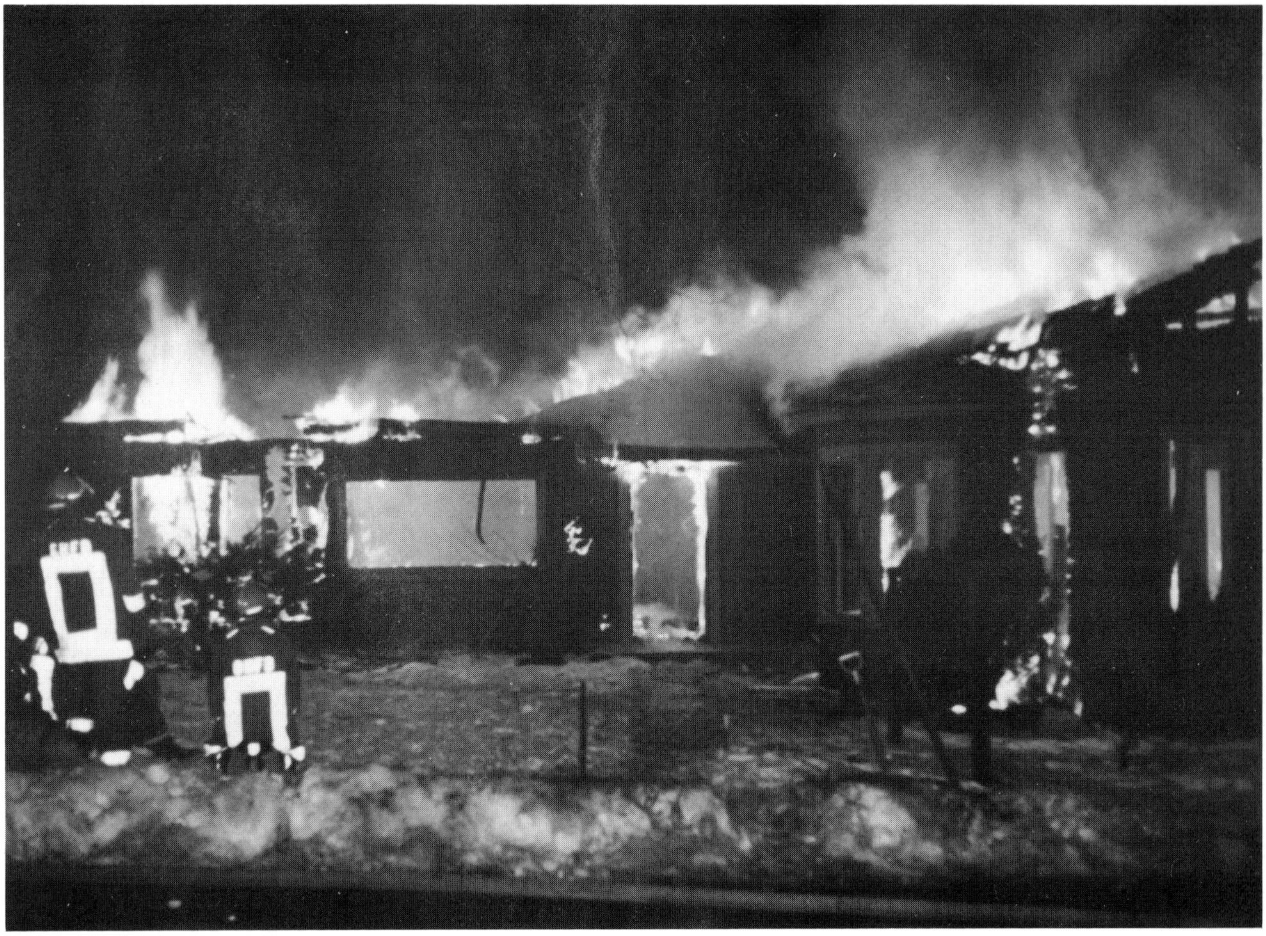

Fully Involved
Judy Galloway
Syracuse, New York

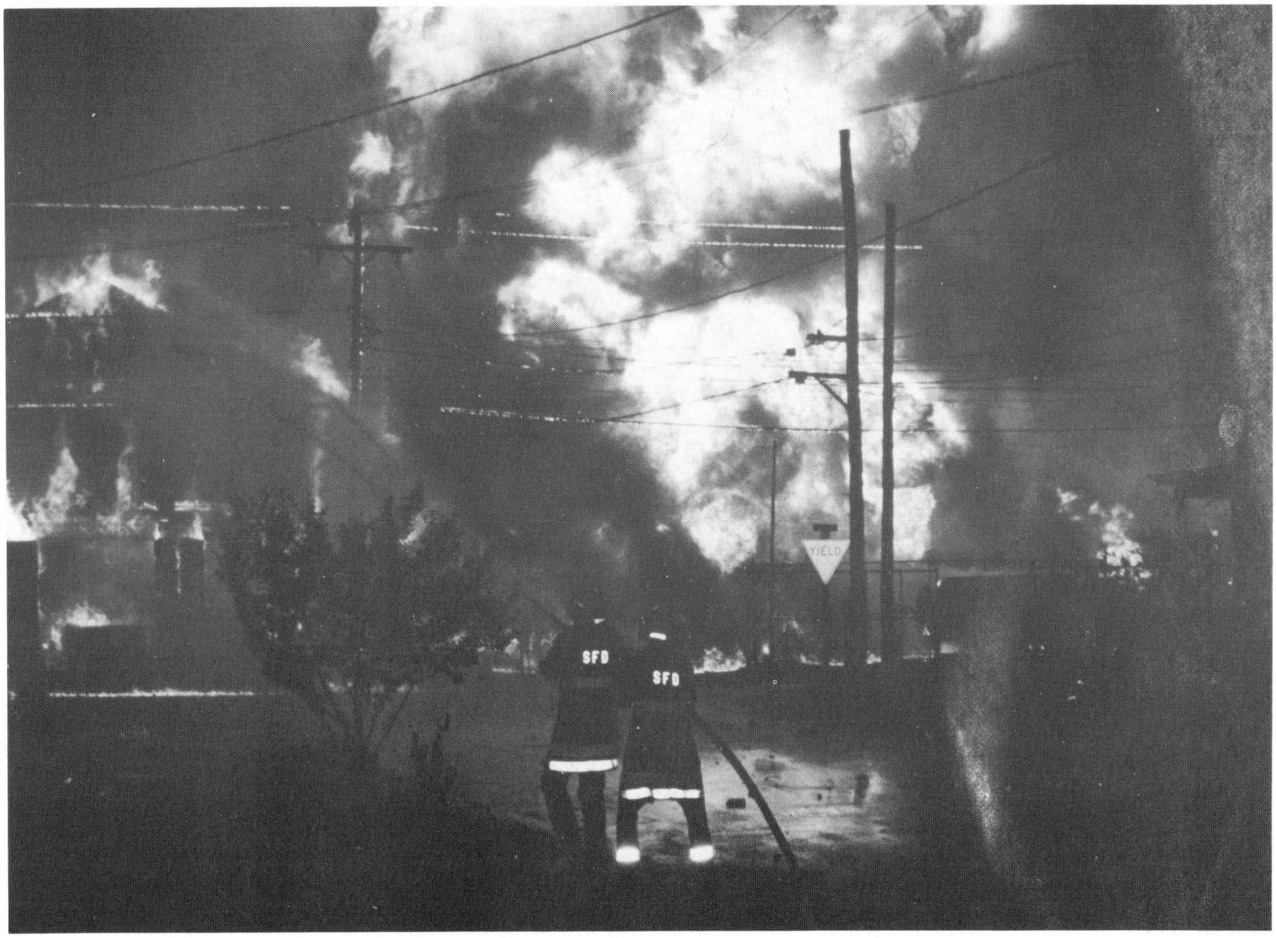

Mirabito Fuel Dump
Frank Provenzon
Sidney, New York

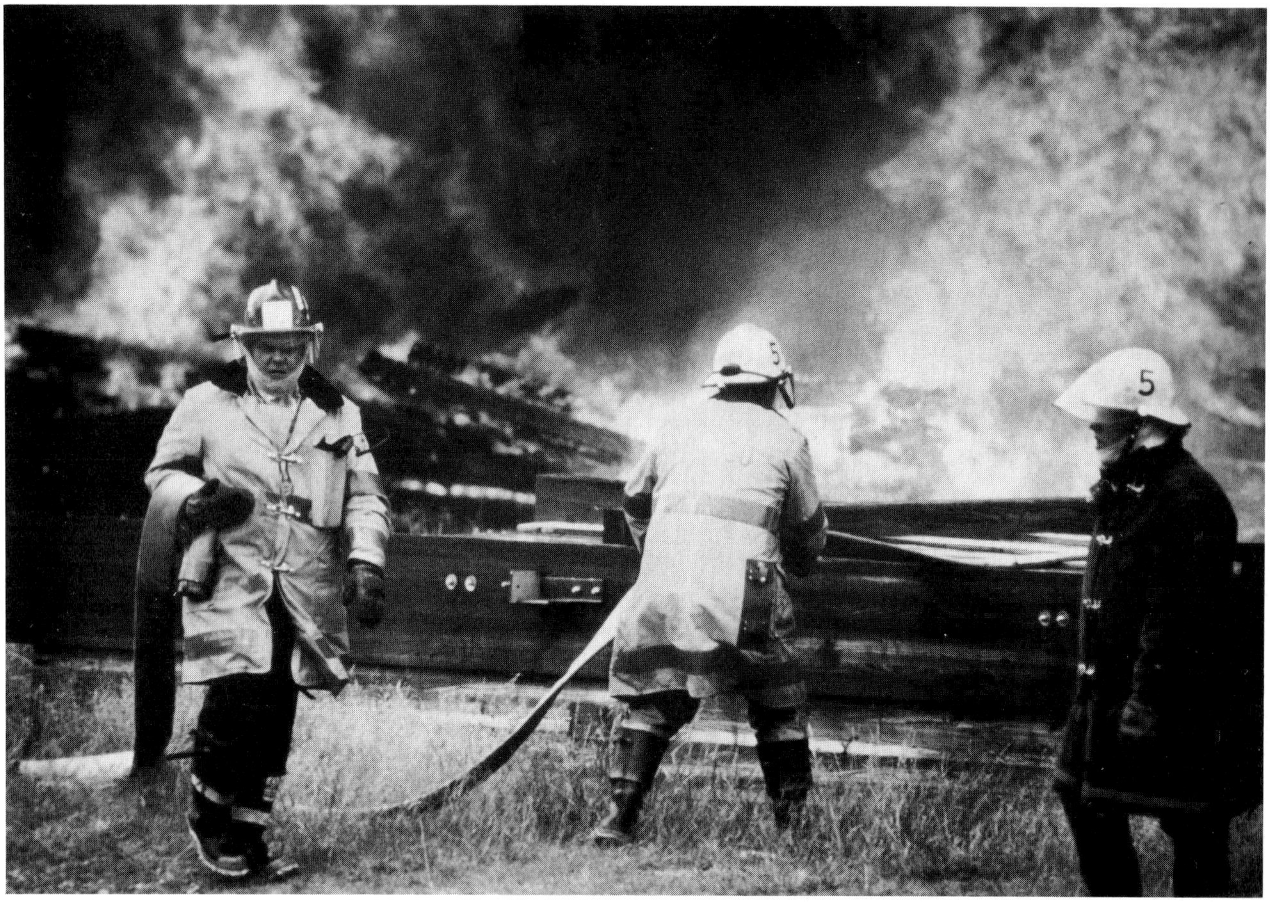

Tom Herde
Lambertville, New Jersey

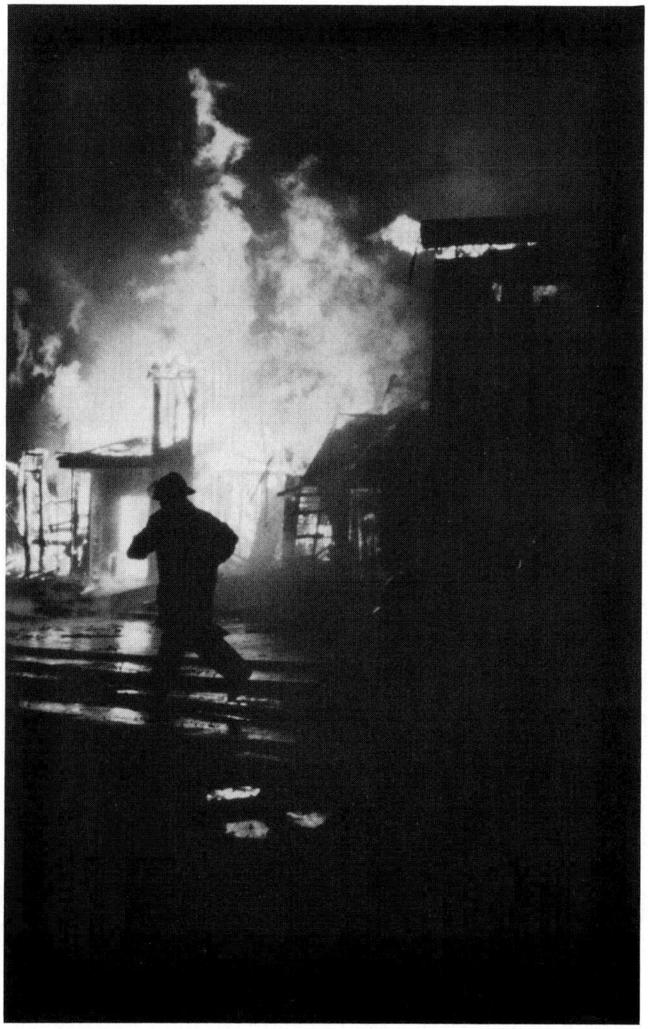

Hustle
John R. Billingham
Garland, Texas

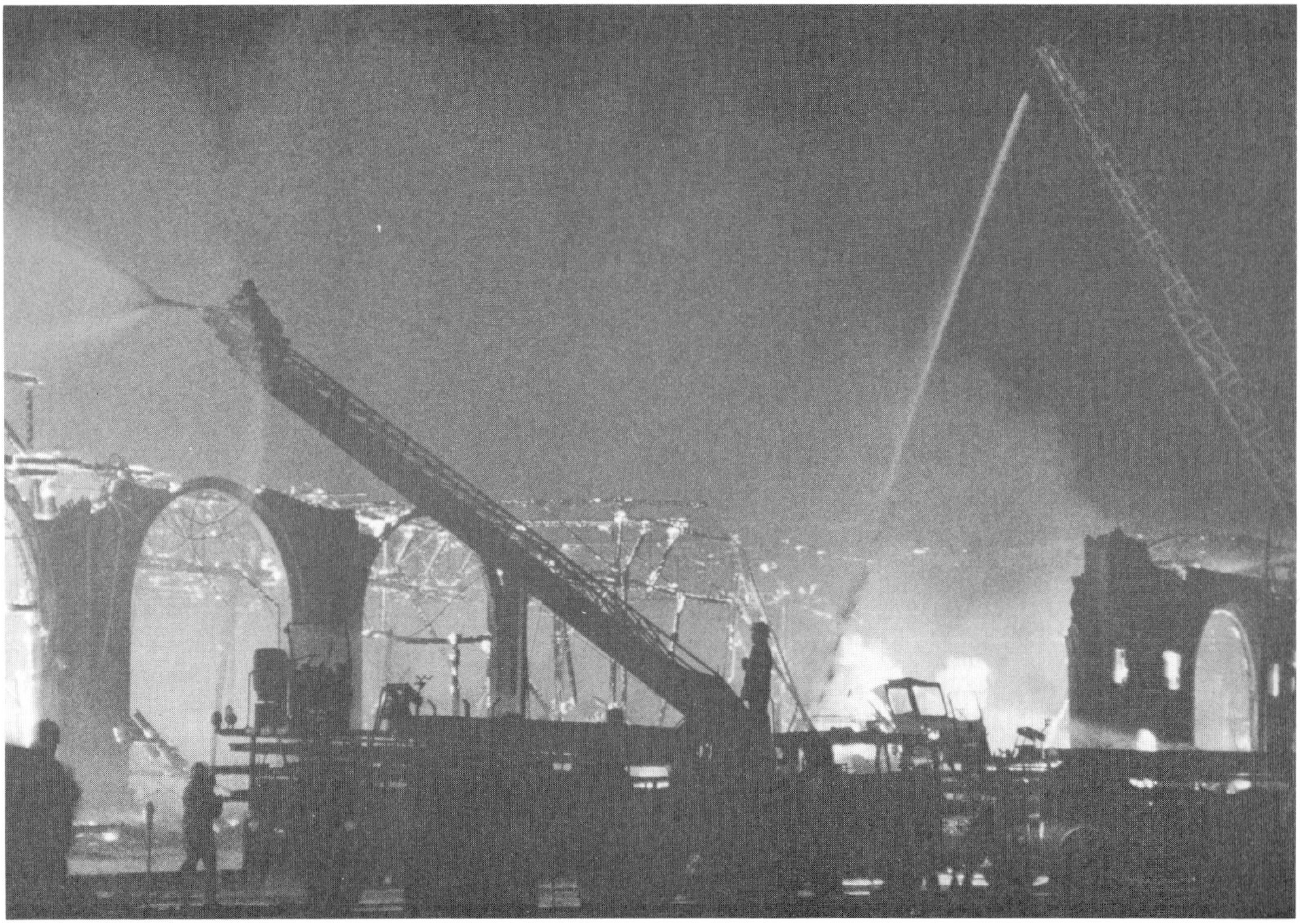

Untitled
Milton Yuen
San Francisco, California

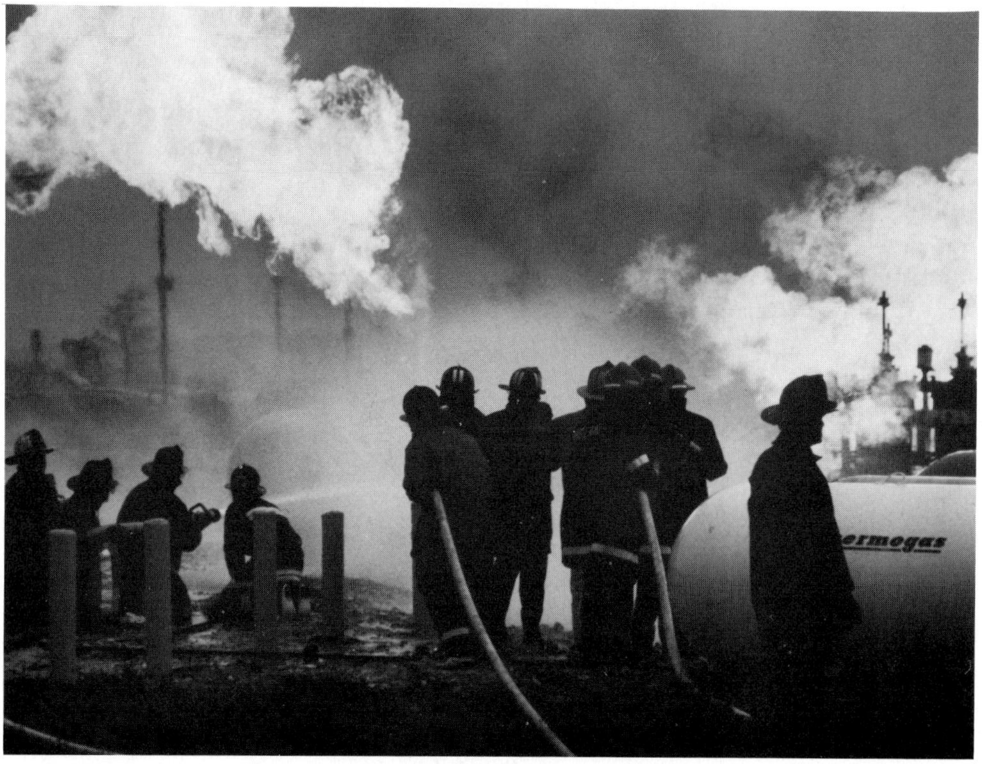

LP - Gas Incident
Ted & Peggy Pegels
Wausau, Wisconsin

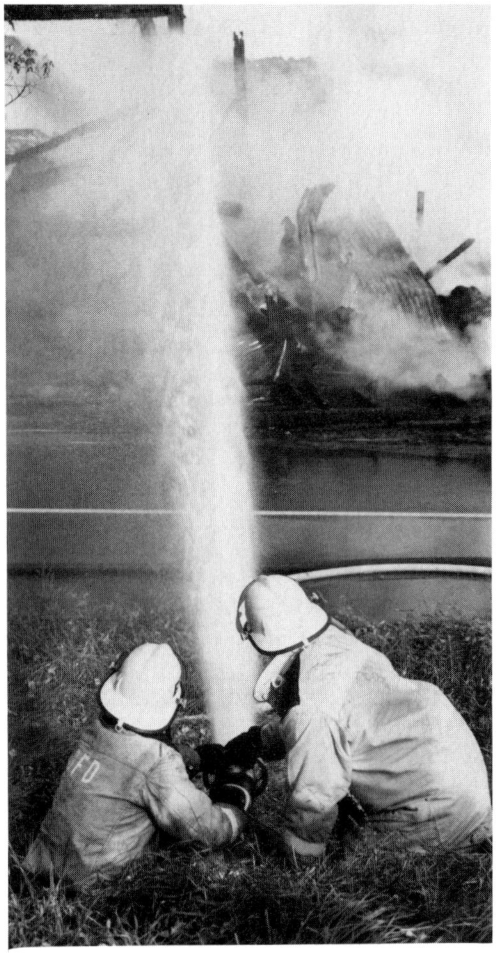

David Seiffert
Granby, Massachusetts

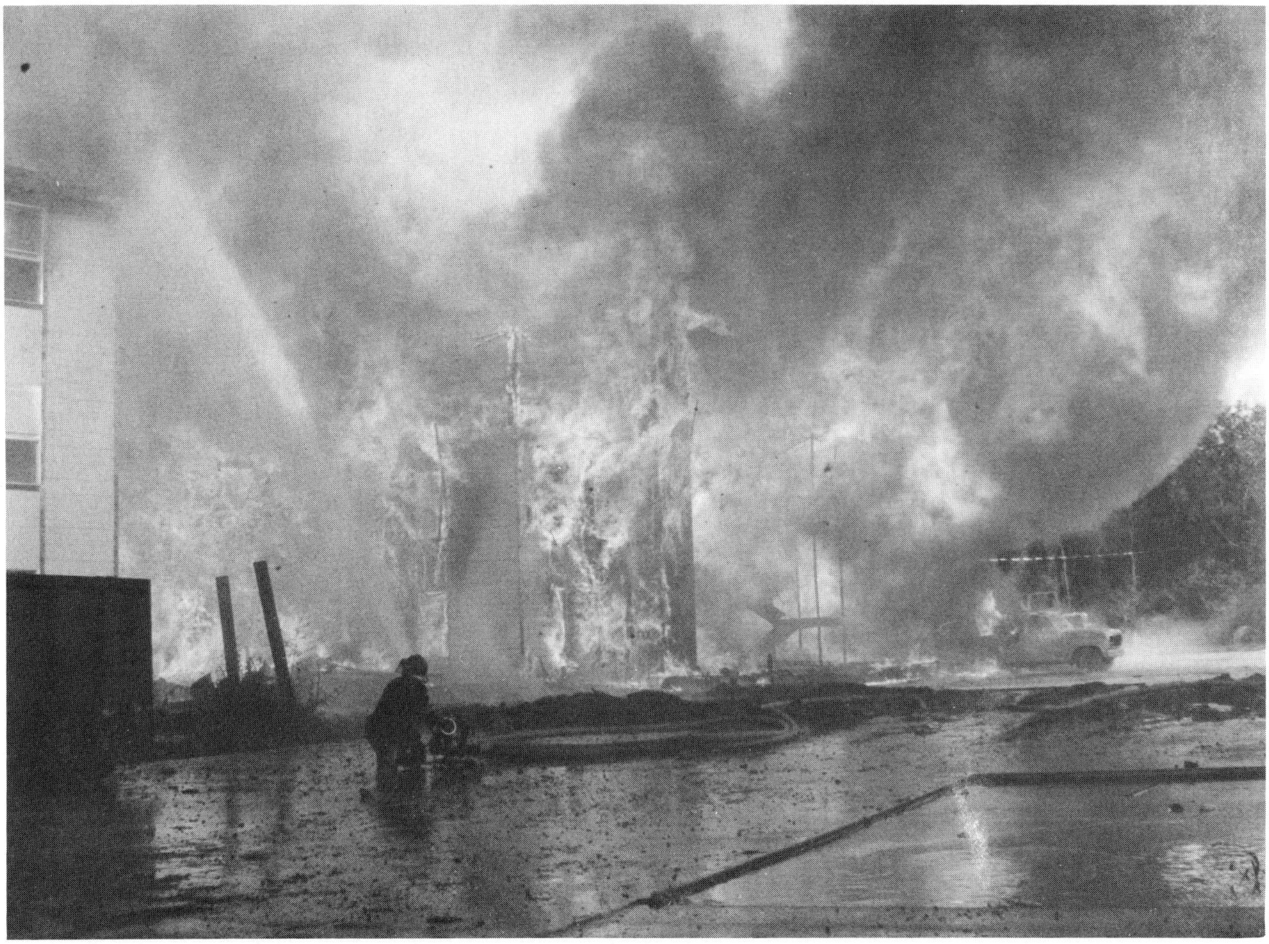

Brian D. Pfister
New Orleans, Louisiana

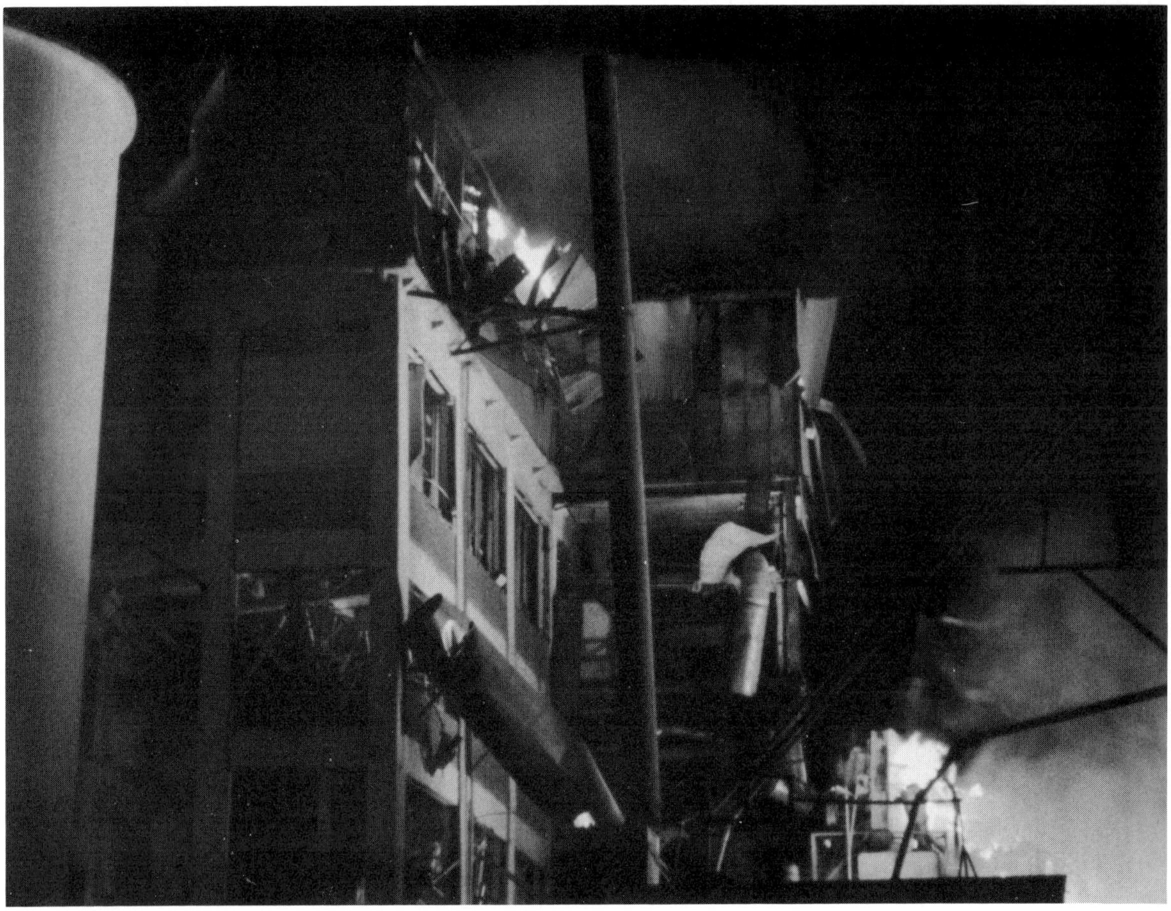

What A Way To
Make Pancake Flour
R.D. Gordon
Garfield, New Jersey

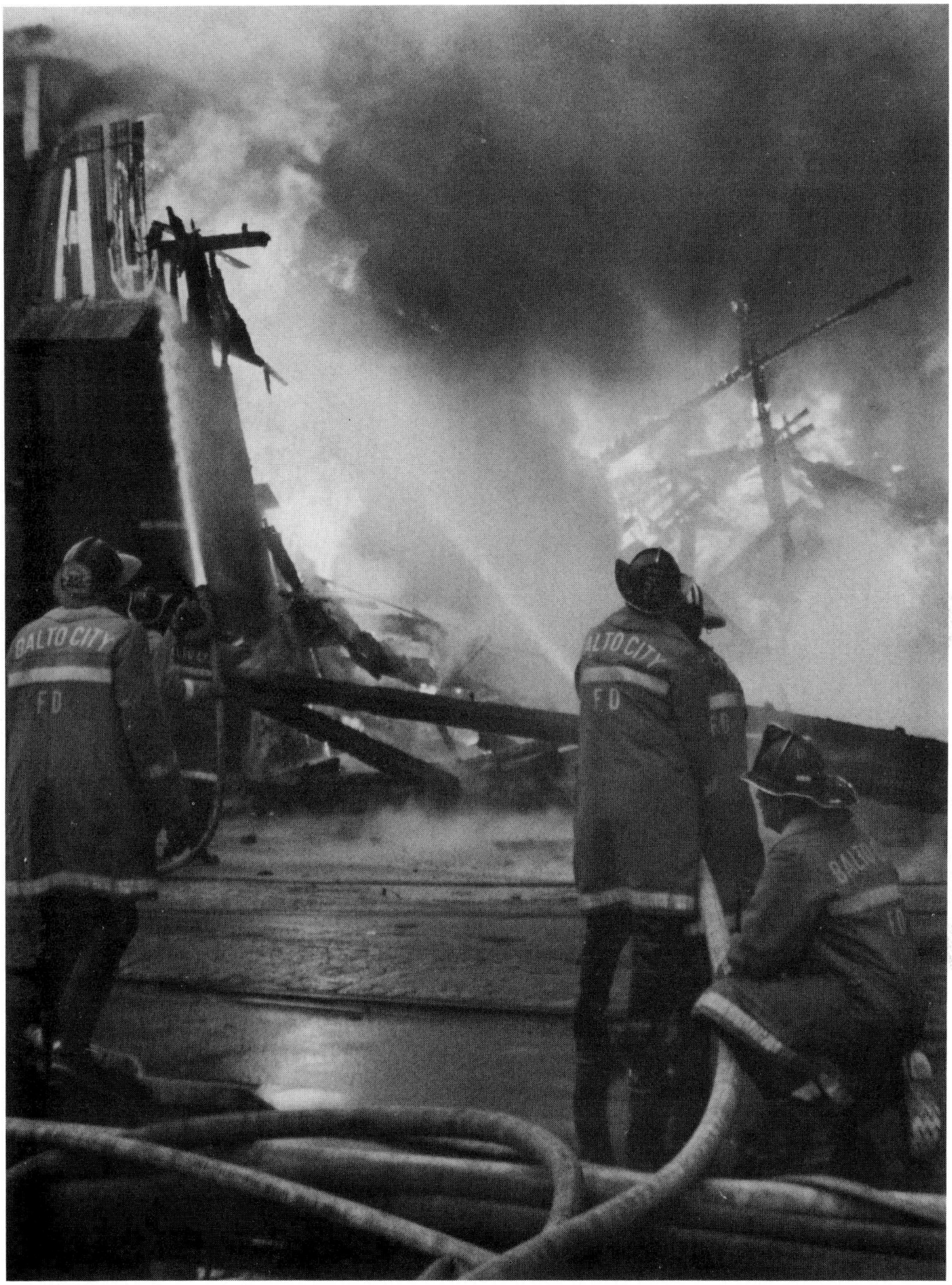

Hot Stand Off
Jim Atkinson
Baltimore, Maryland

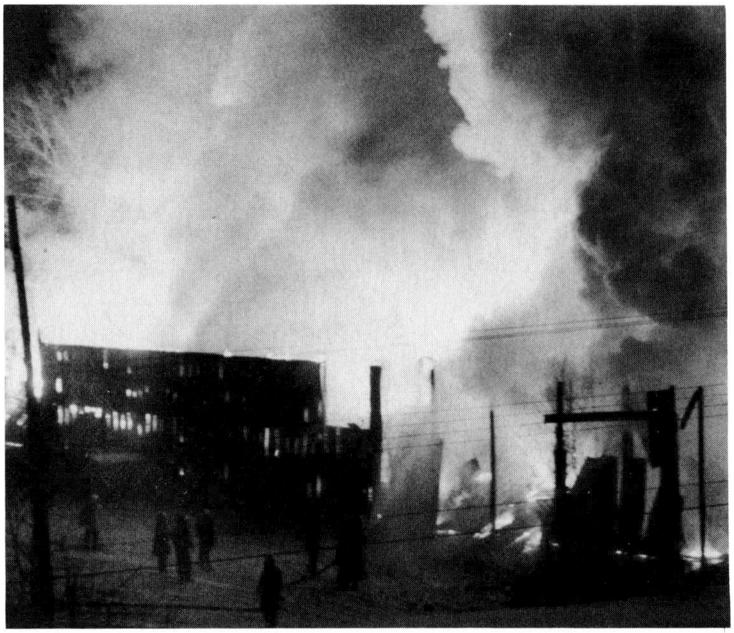

In the Dawn's Early Light
Chief Greg T. Lemons
Cynthiana, Kentucky

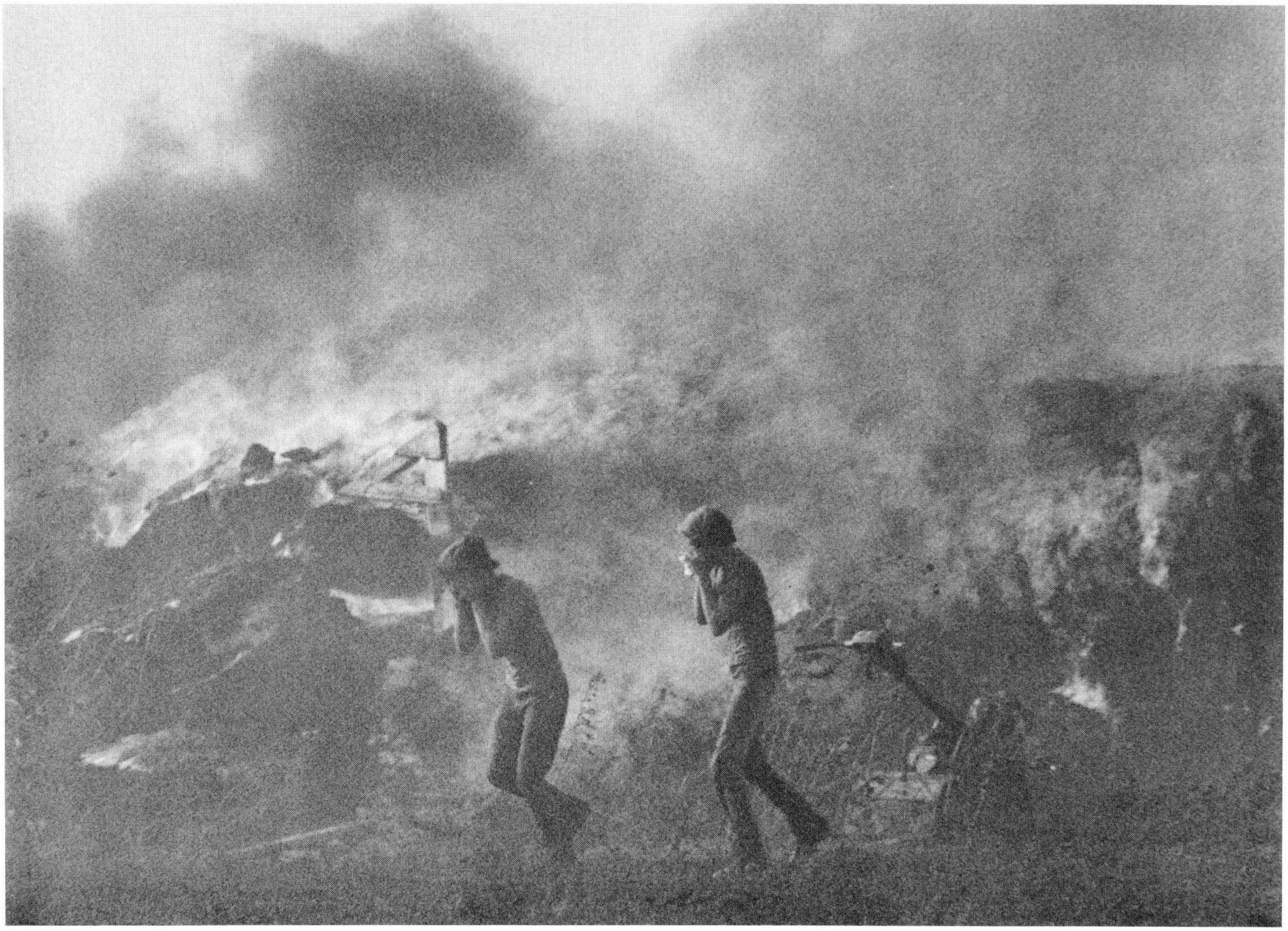

Running for Safety
Larry R. Fleming/Wichita Eagle-Beacon
Wichita, Kansas

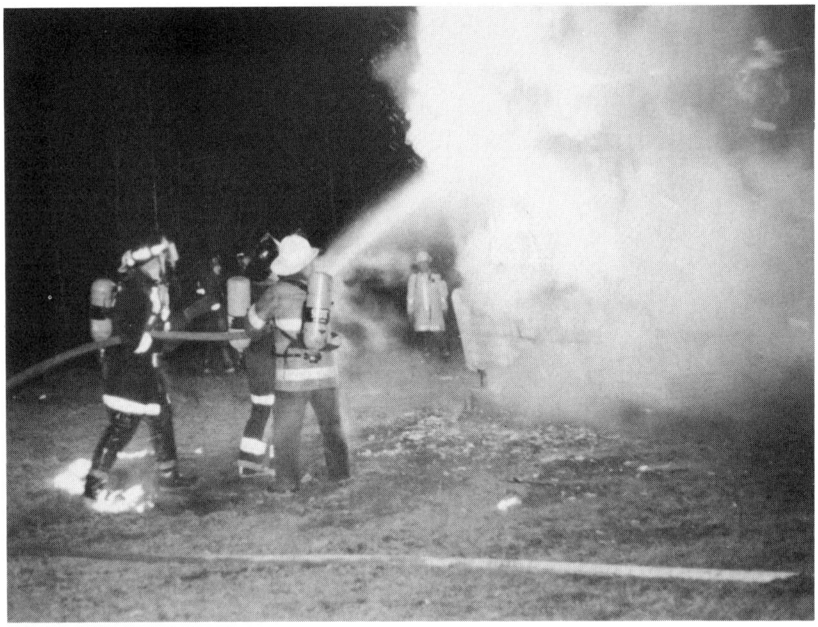

Homer Walters
Wooster, Ohio

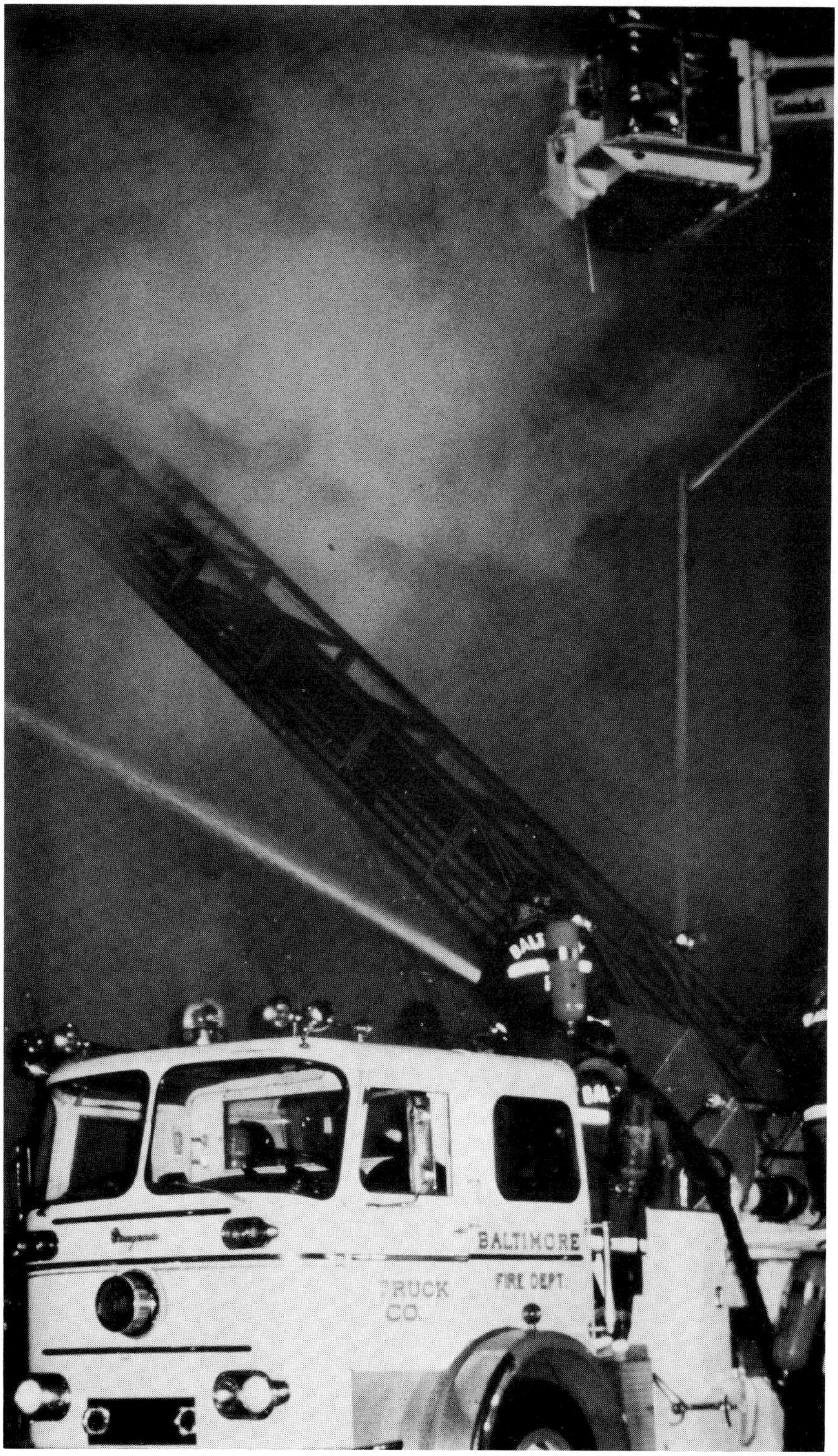

Baltimore Firefighting
Jim Atkinson
Baltimore, Maryland

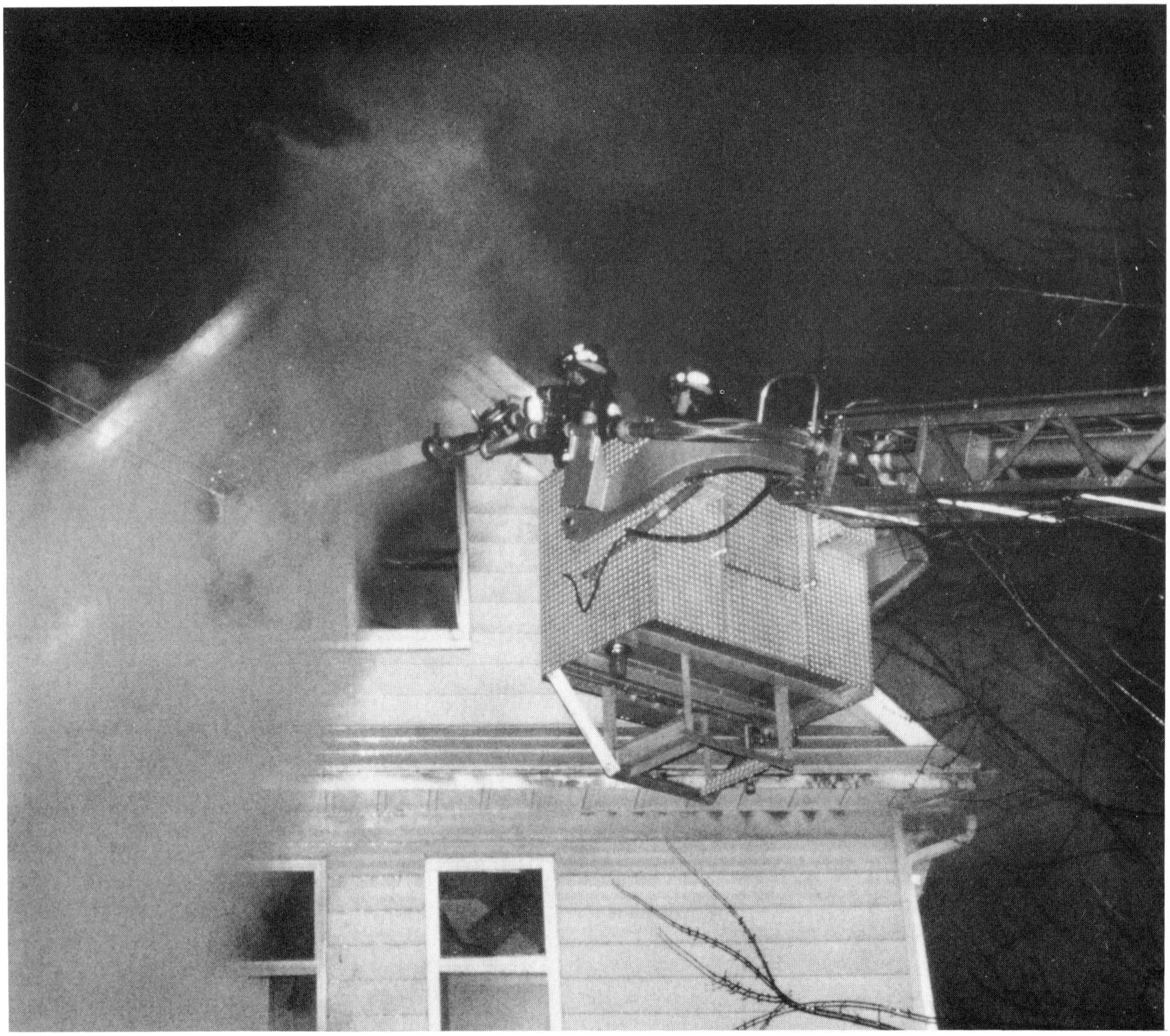

Opening Up
Peter A. Danzo
Ridgefield Park, New Jersey

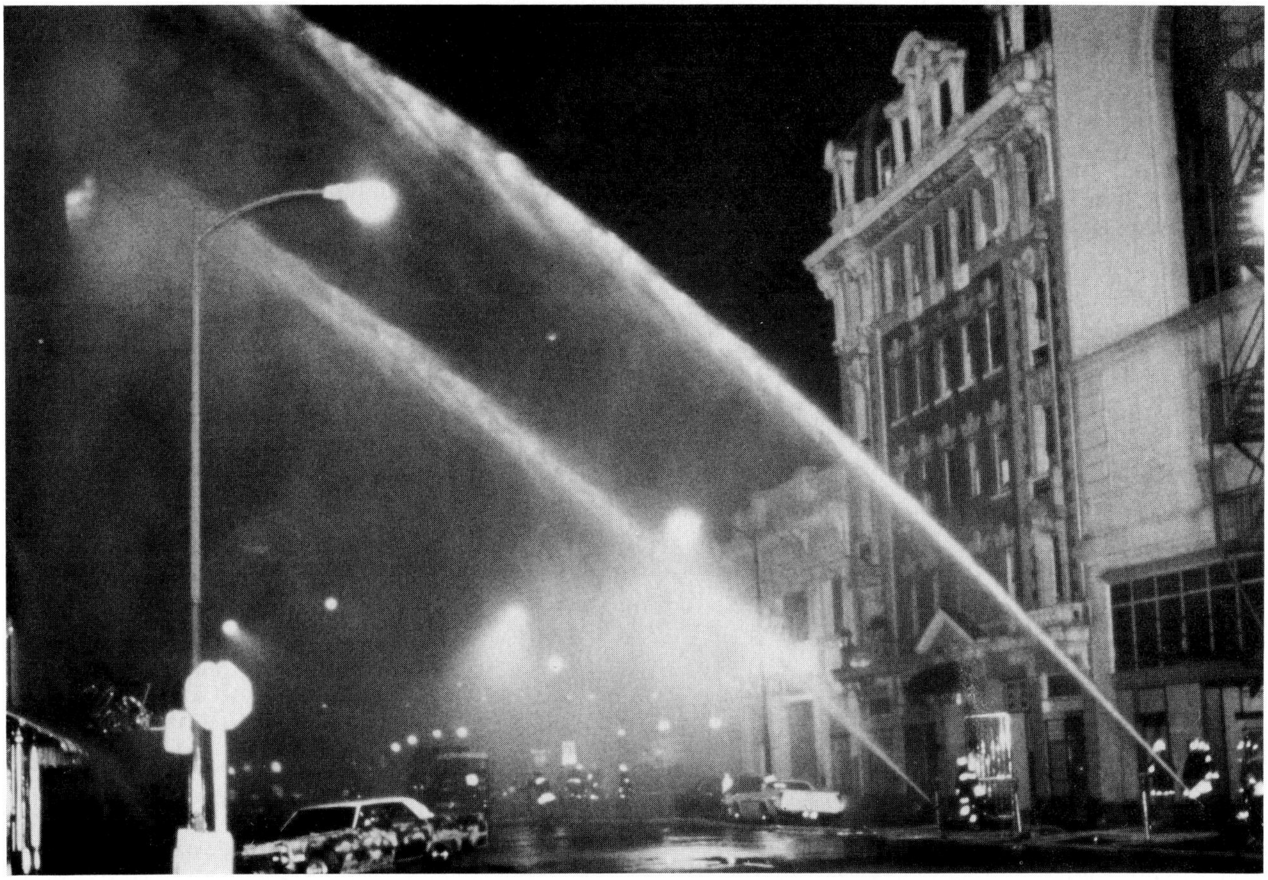

Clark Hotel
Richard Cowan
Stockton, California

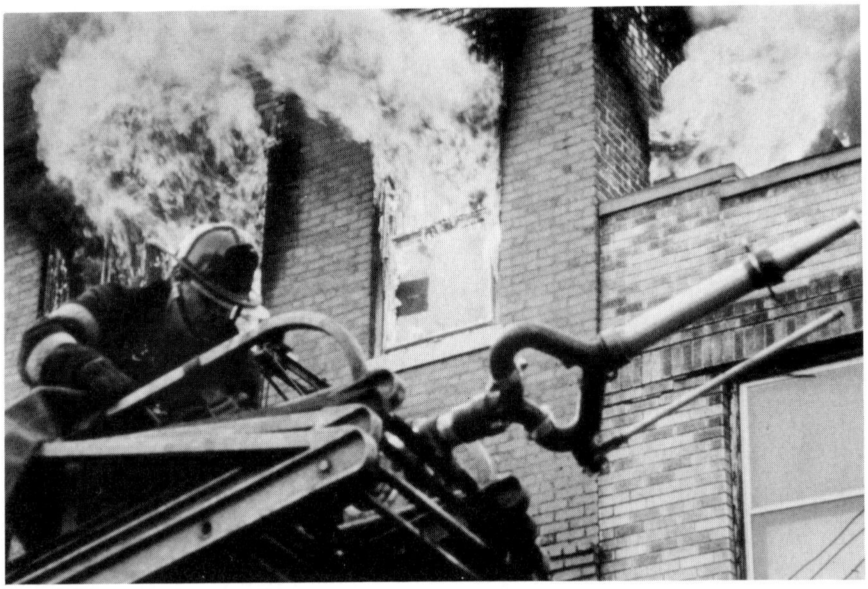

It's Ladder Pipe Time
Ronald Jeffers
Union City, New Jersey

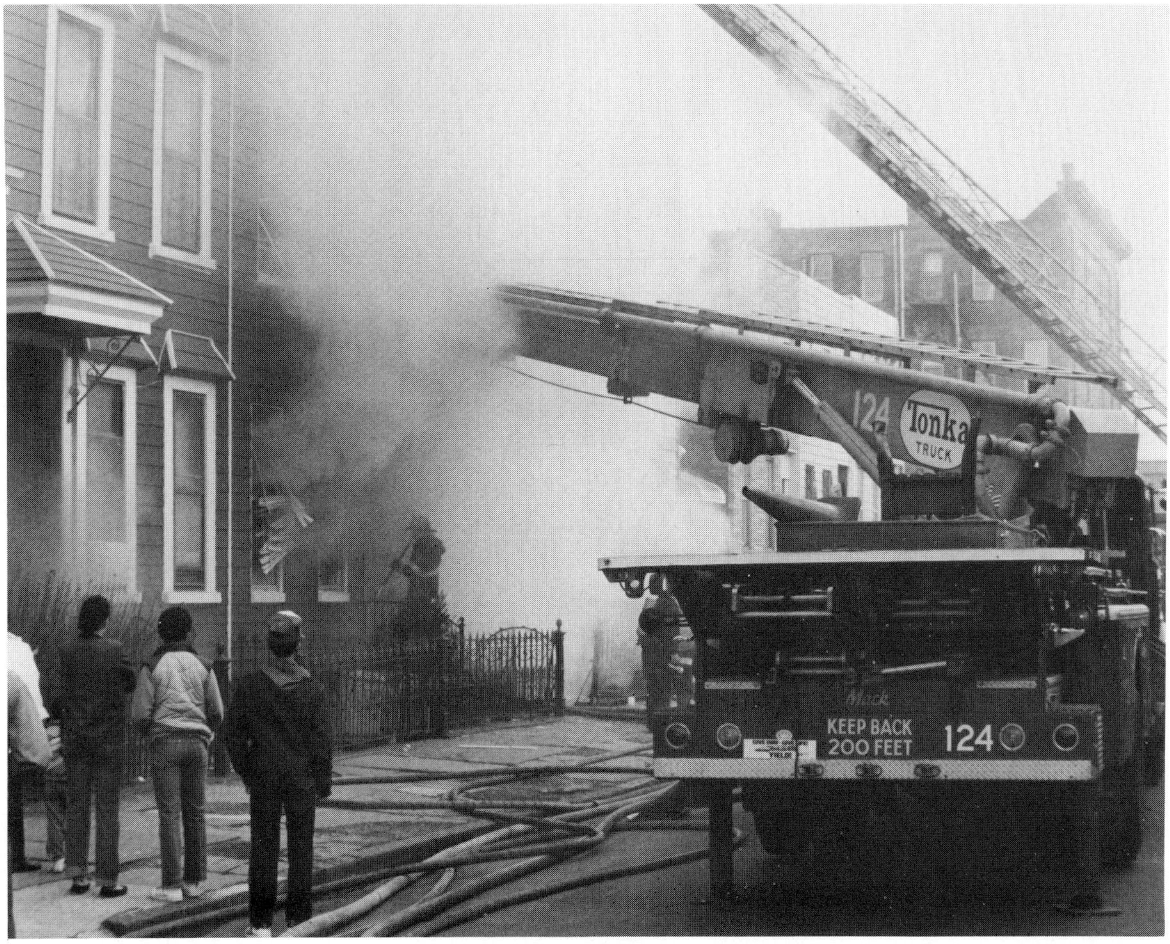

A Real "Snotty" Job
James Sanquedolce
Mystic, Connecticut

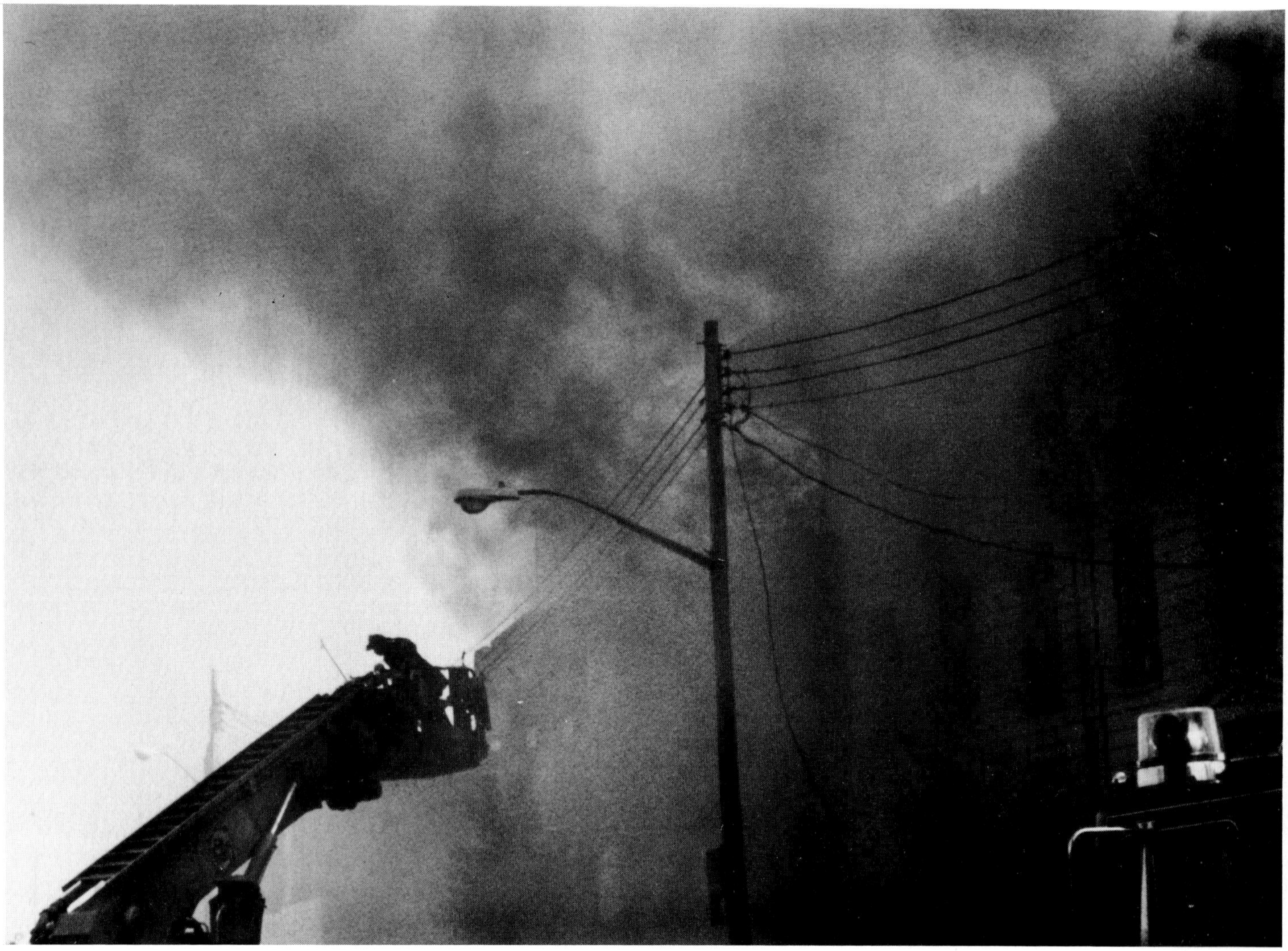

3rd Alarm
Ed Heavey
Deer Park, New York

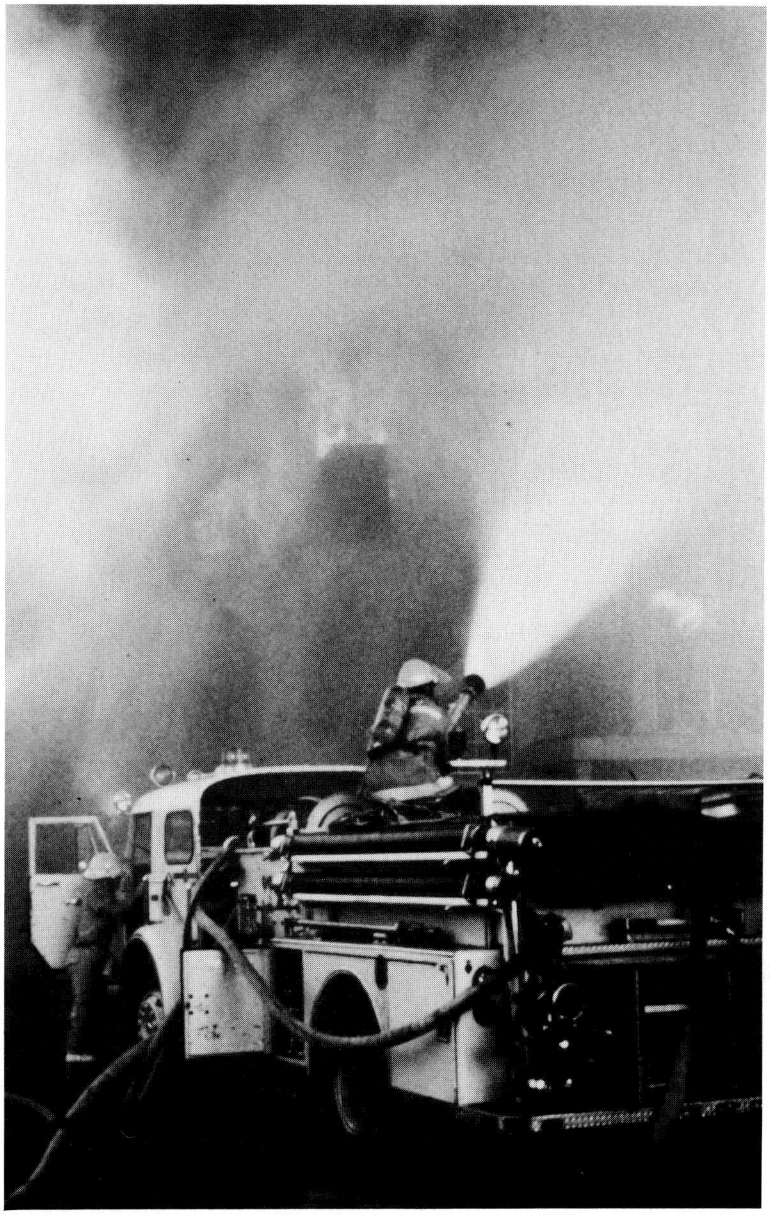

David White
College Station, Texas

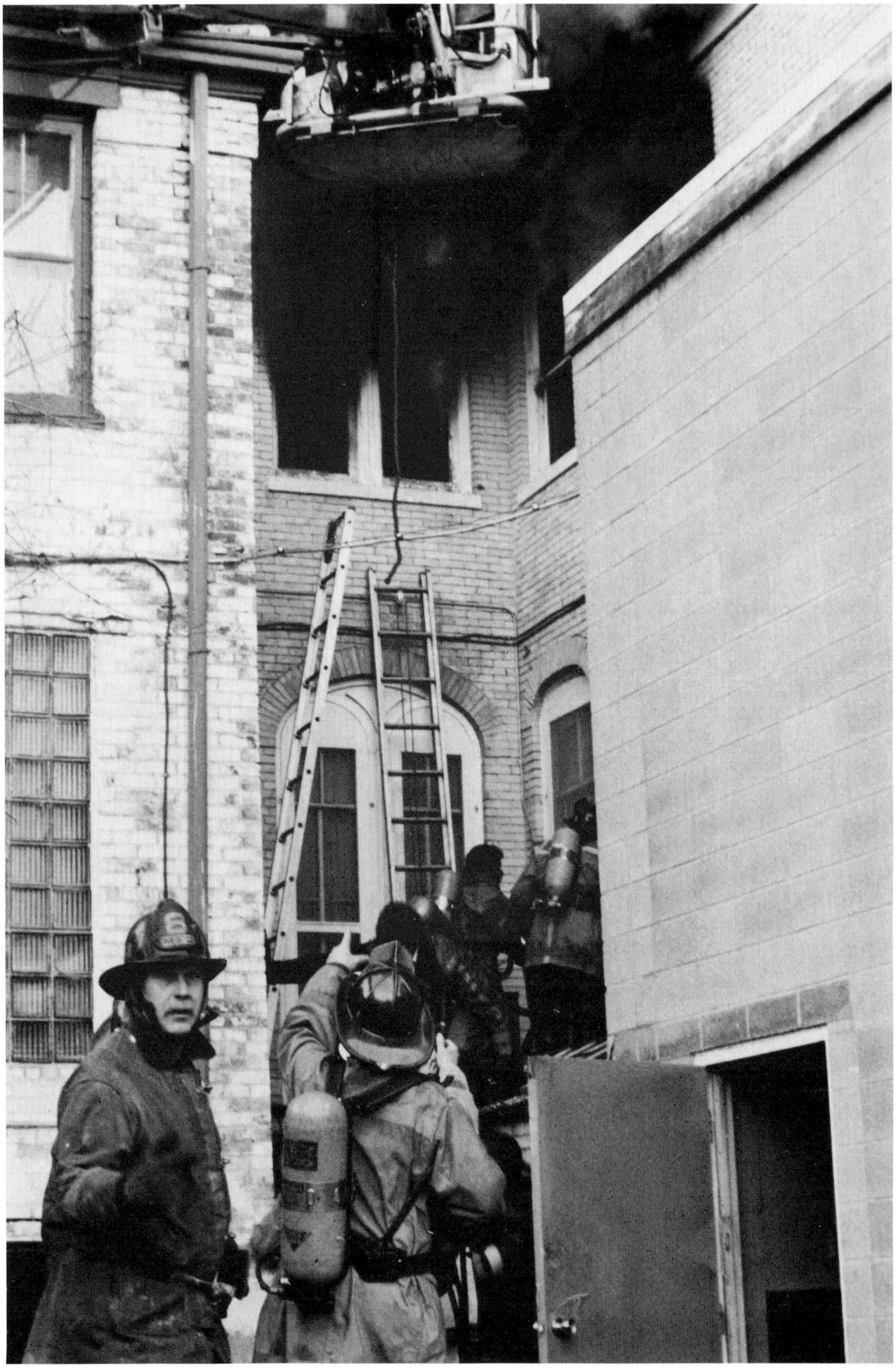

Just In Time
Mark Hoeller
Milwaukee, Wisconsin

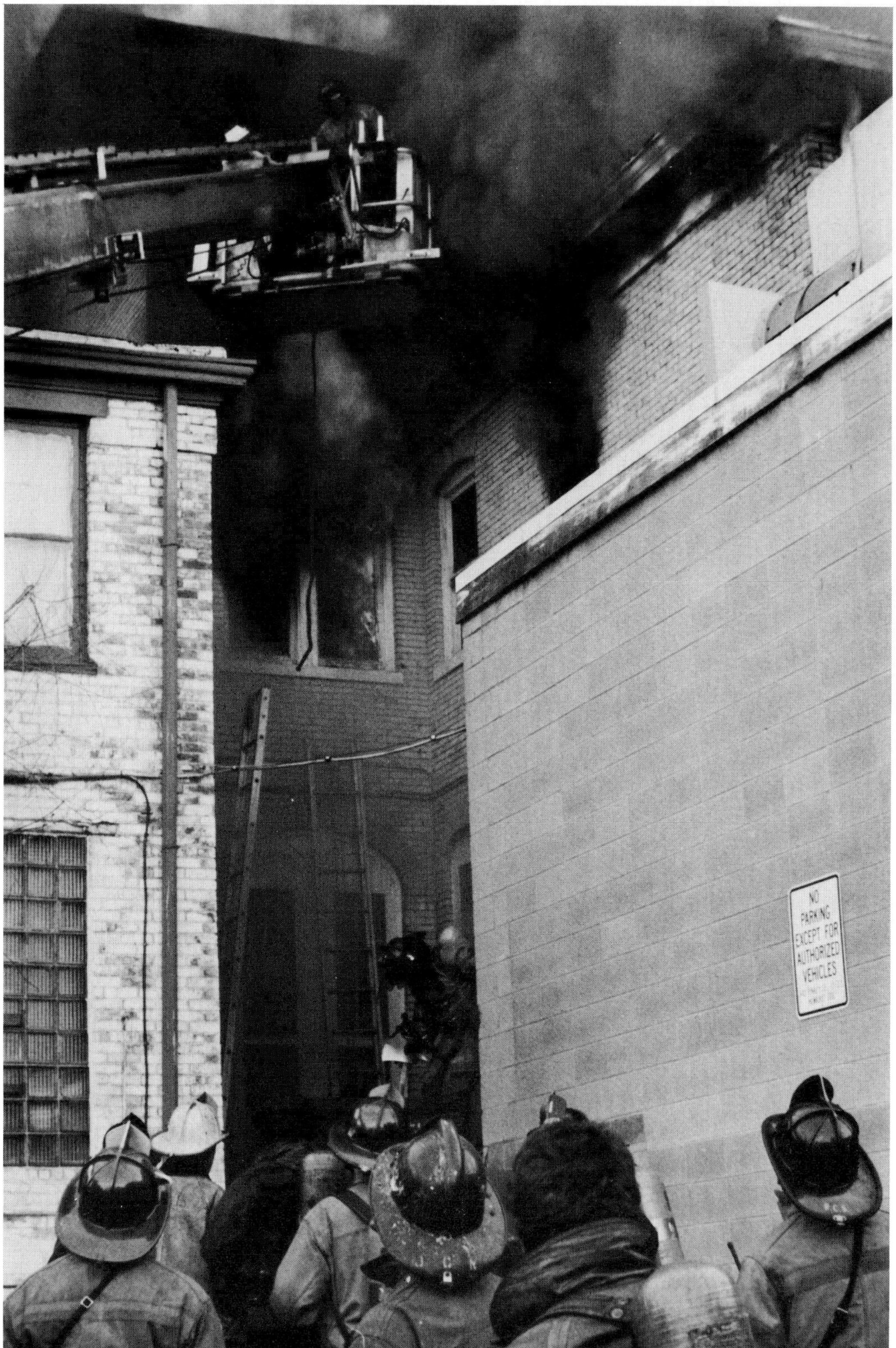

Mark Hoeller
Milwaukee, Wisconsin

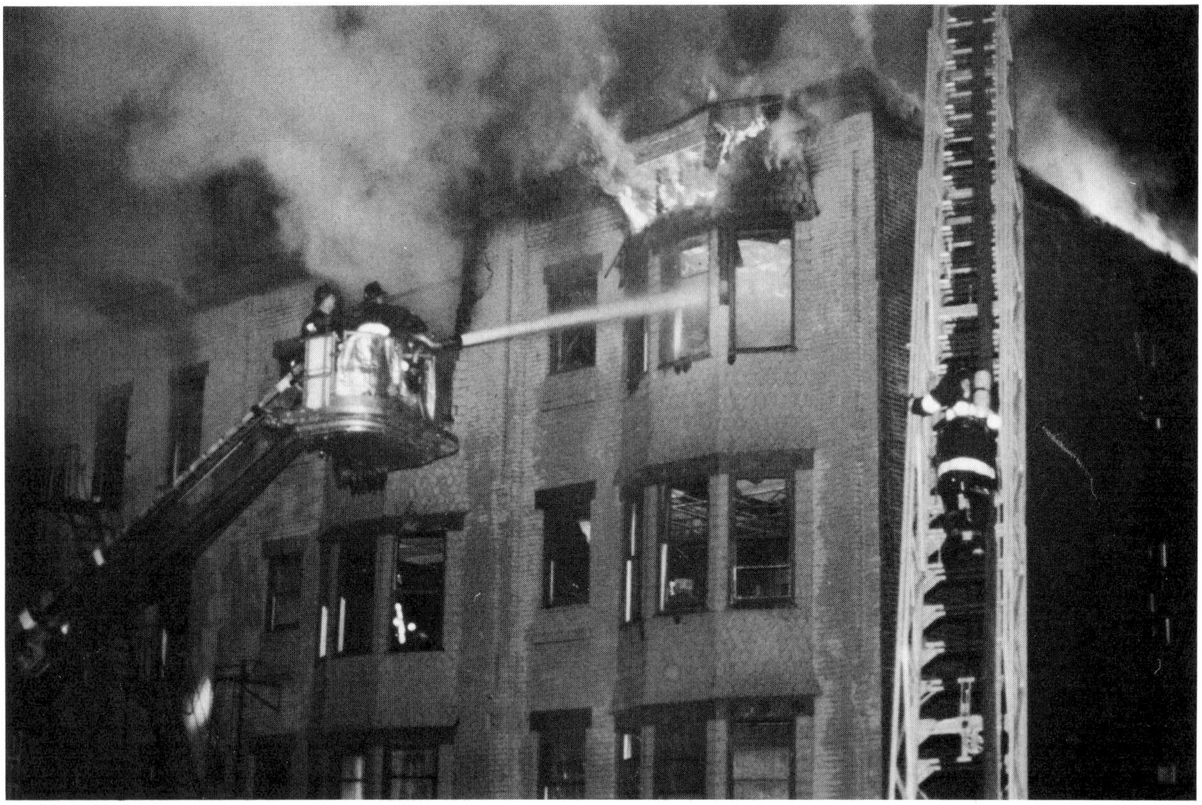

2 Alarm Yonkers, NY
Harold LaRocco
Porchester, New York

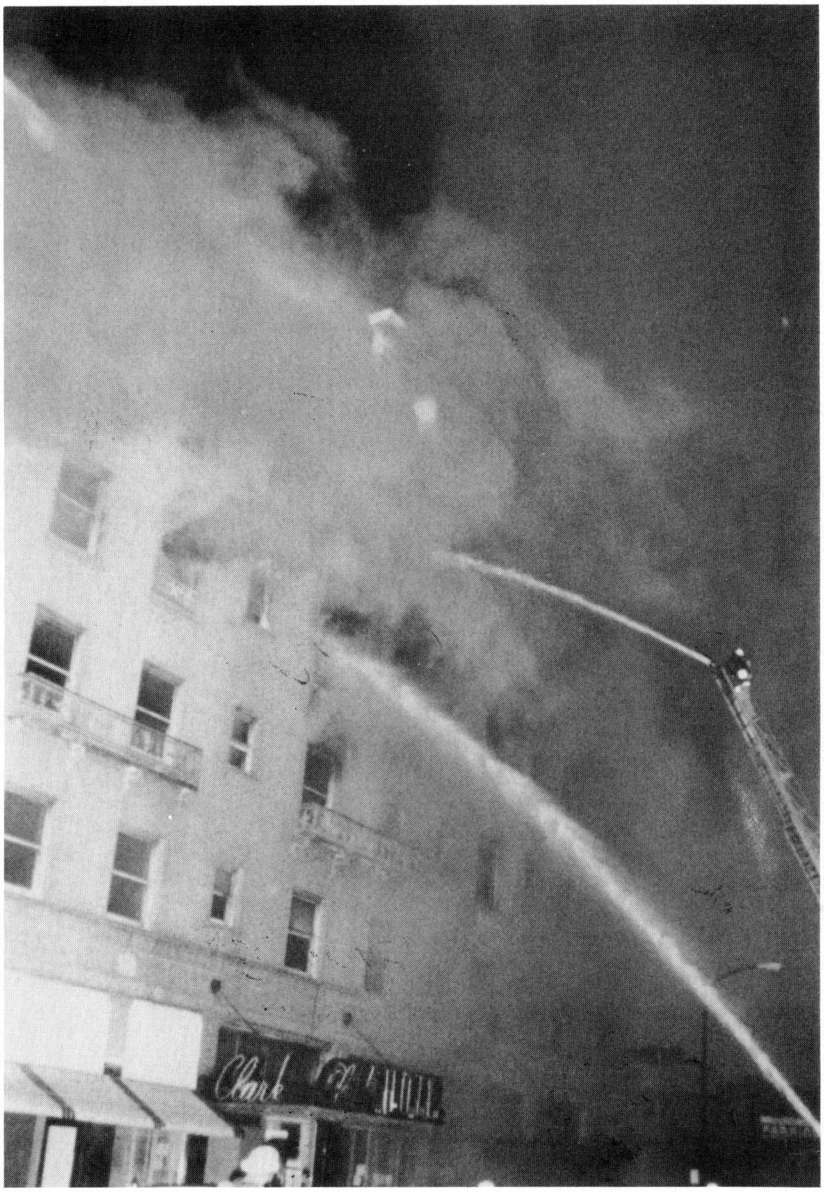

Master Streams From
The Street
Vicki Cowan
Stockton, California

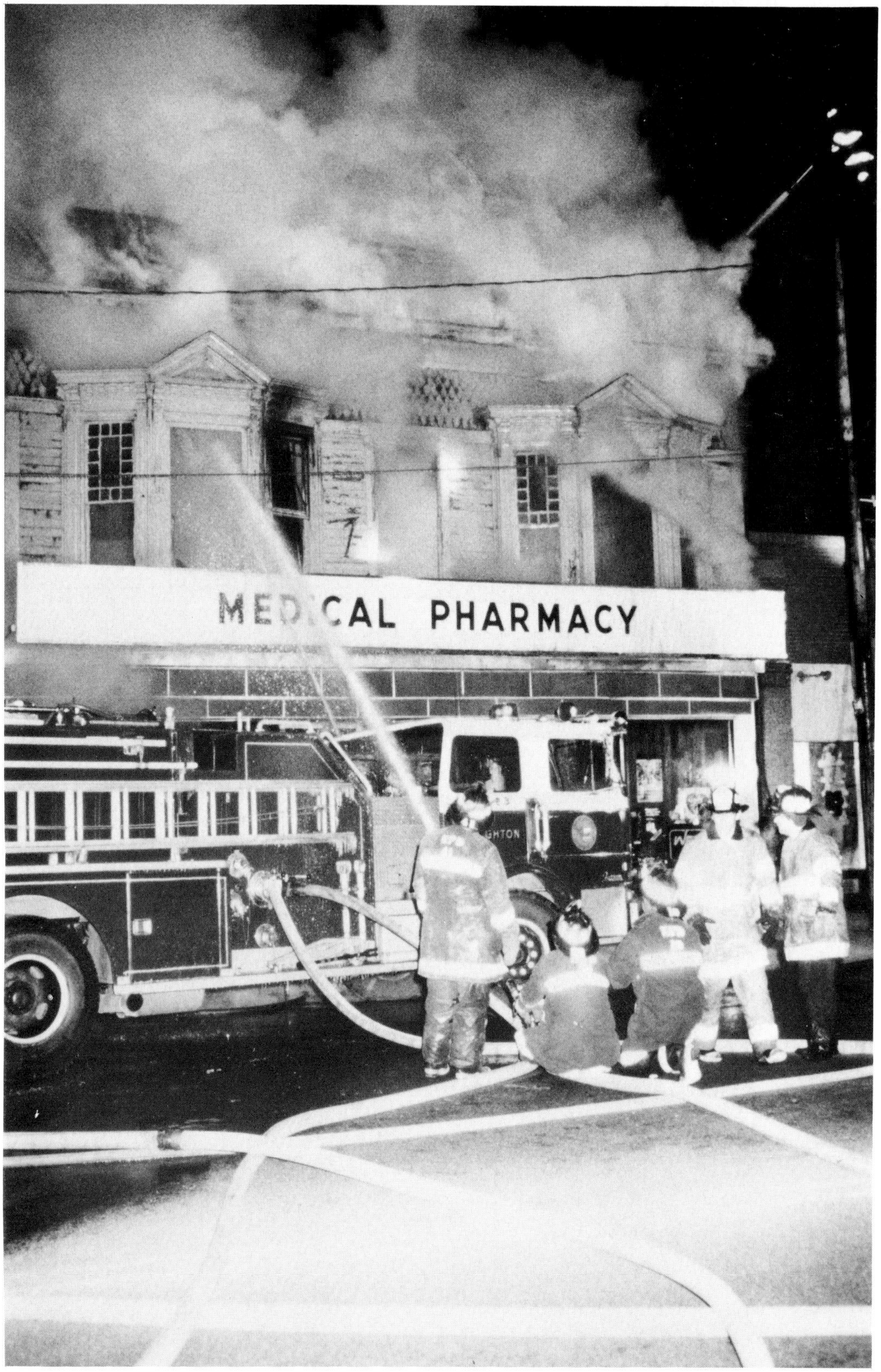

Stoughton Box 3-3
Robert Myers
Brockton, Massachusetts

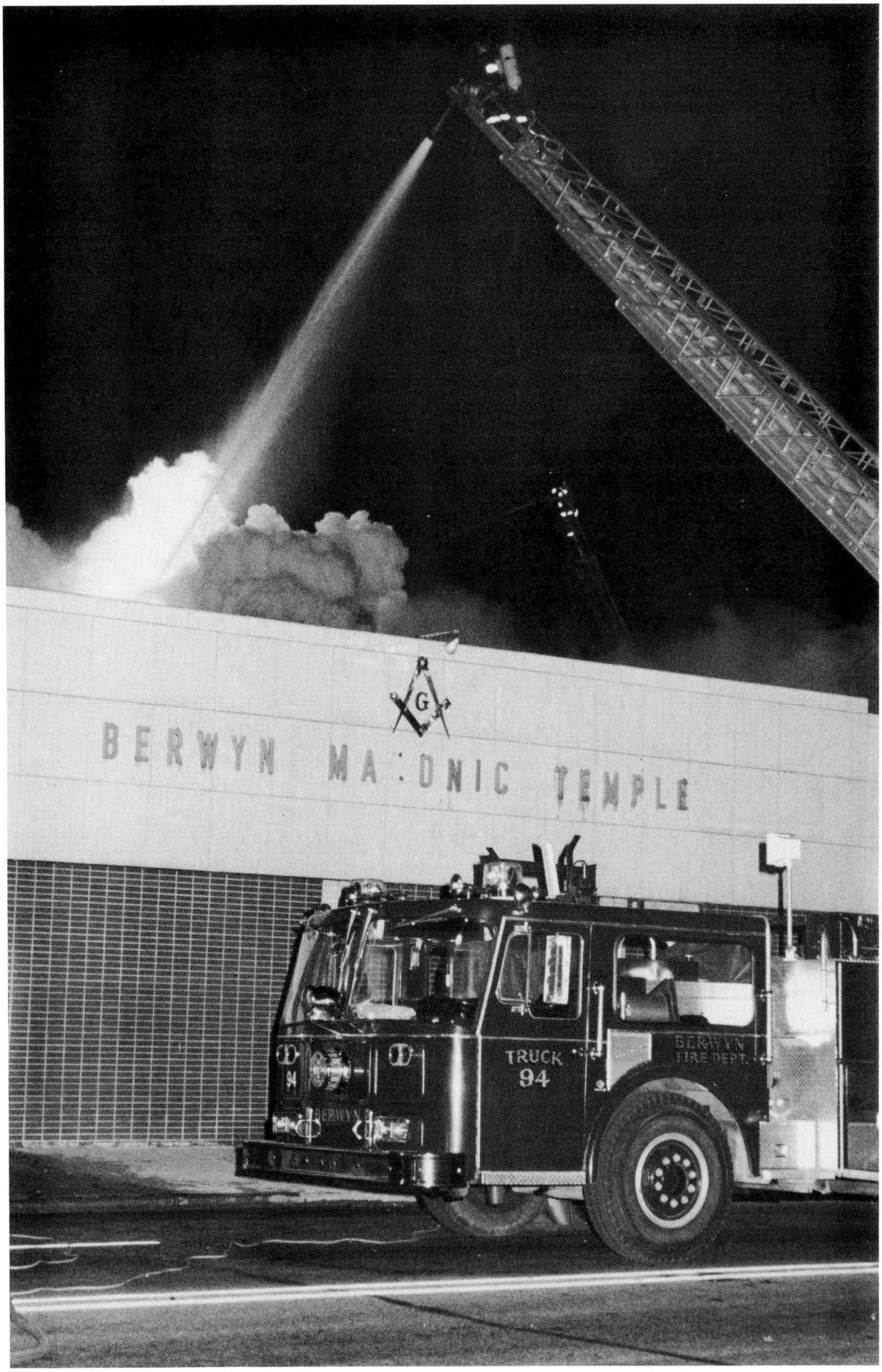

Fire of the Gods
Paul Gardner, Jr.
Berwyn, Illinois

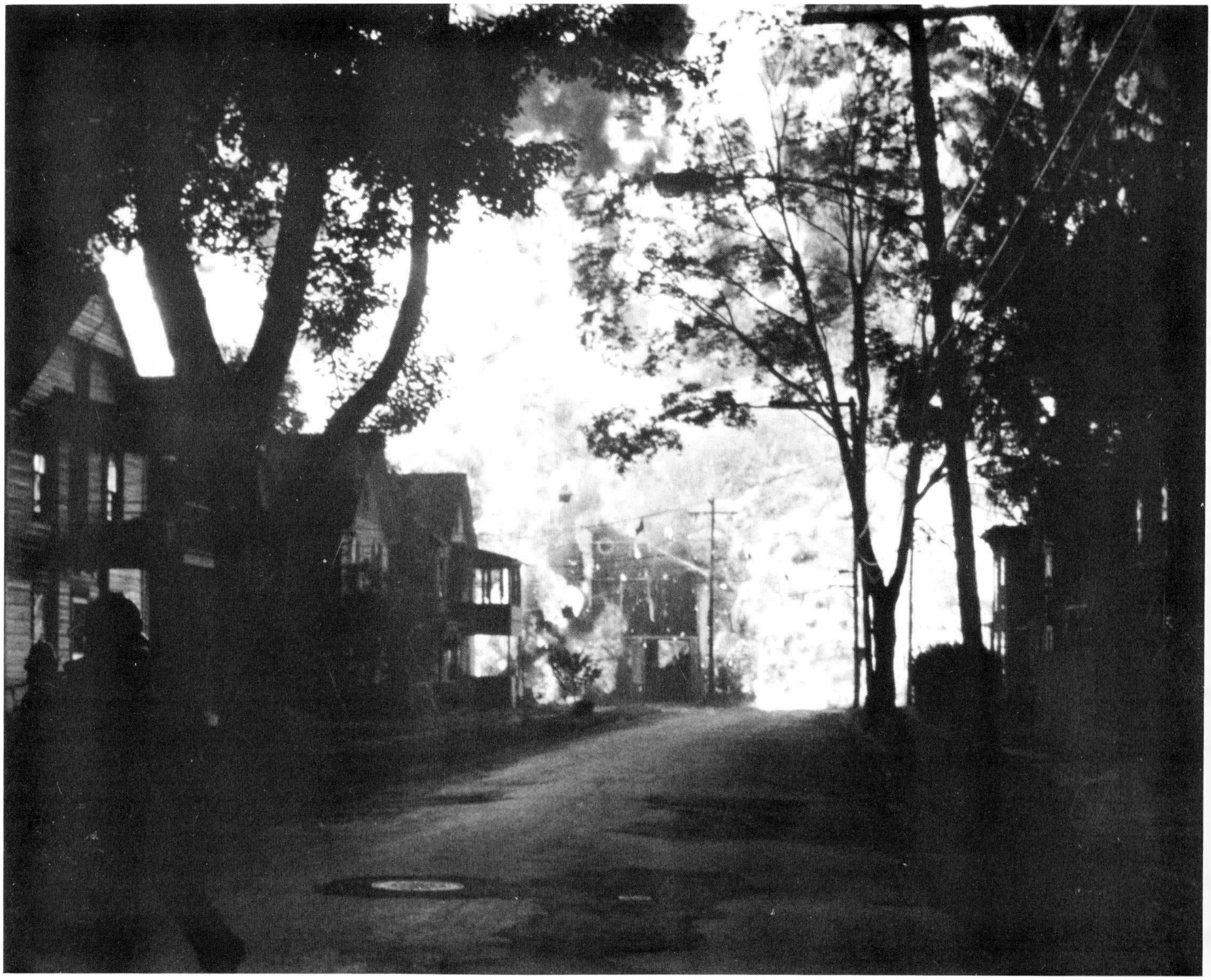

Mirabito Fuel Dump
Frank Provanon
Sidney, New York

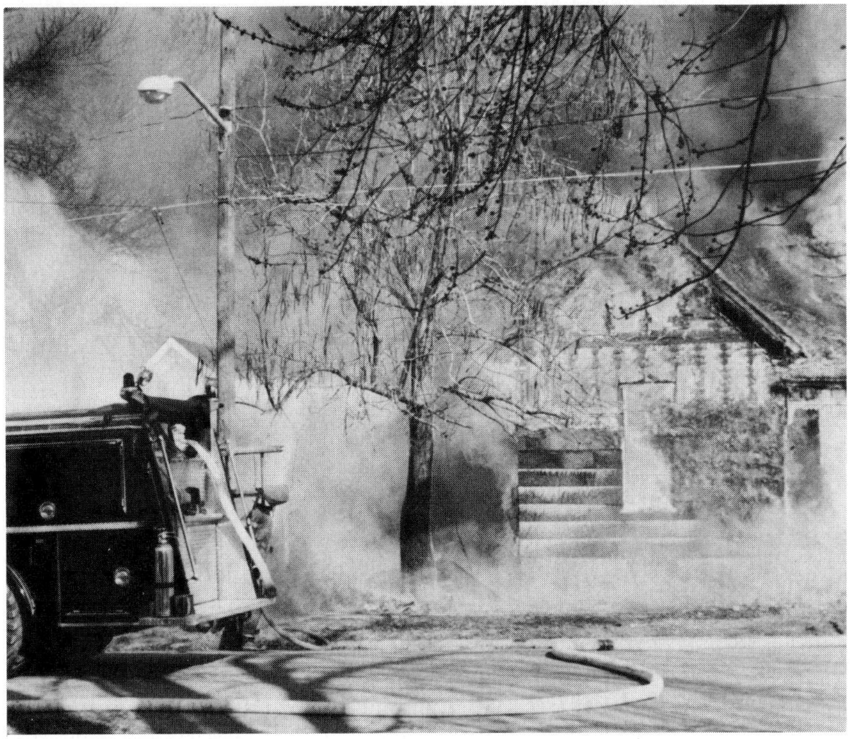

Arson Fire
Rhonda Herron
Marshall, Missouri

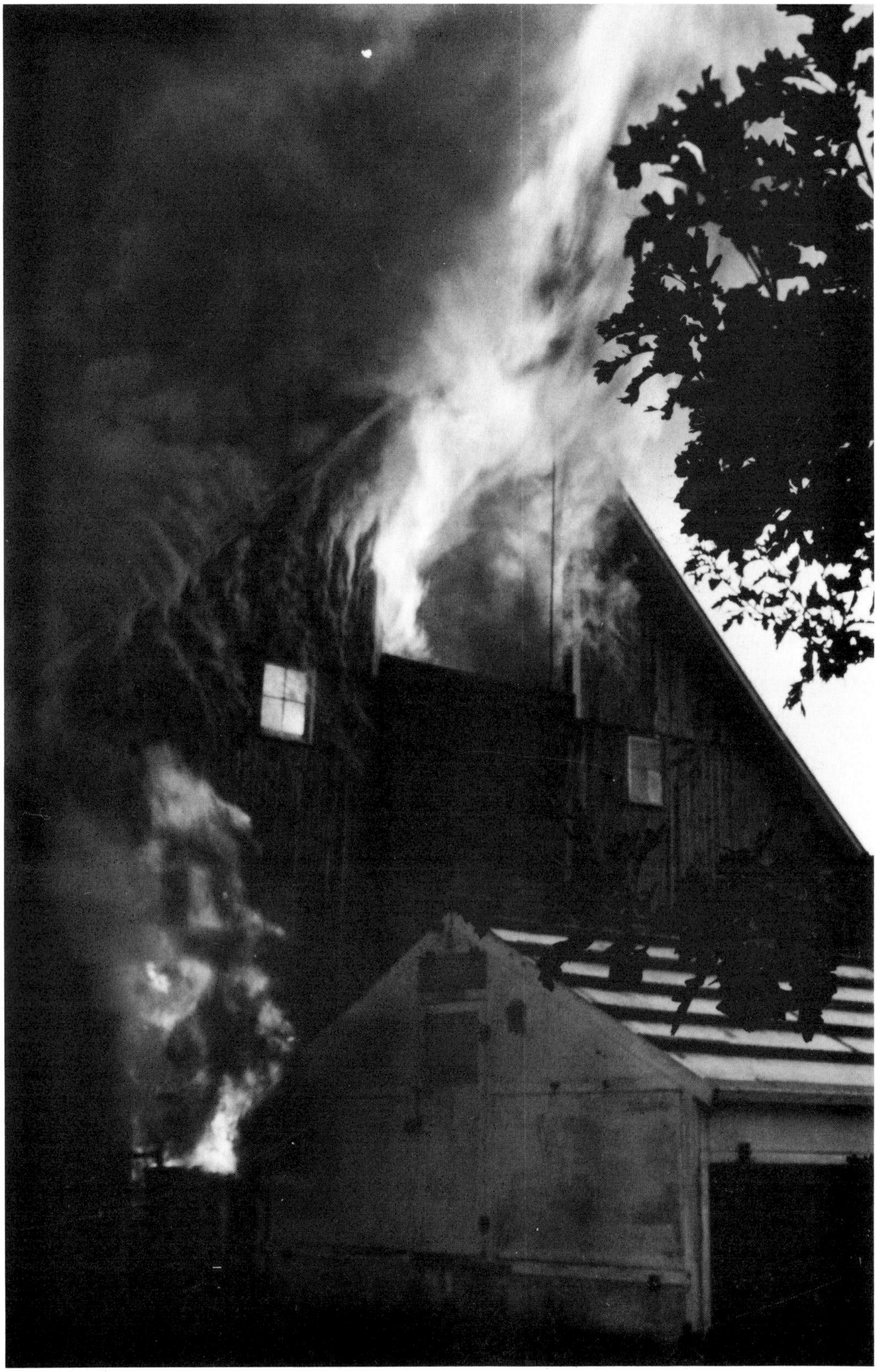

Ernest J. Hansen
Gladbrook, Iowa

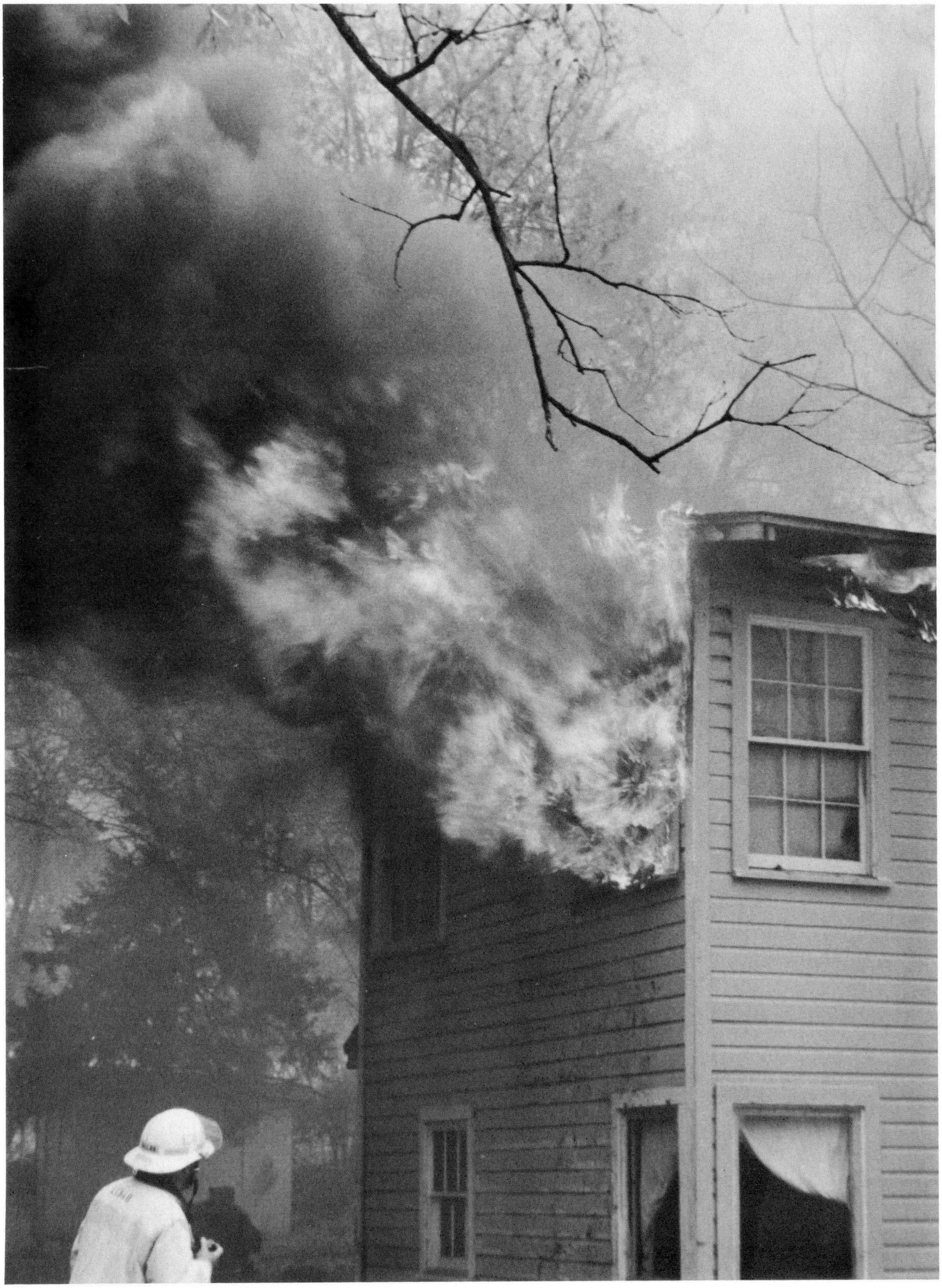

Lots of Fire
Karen Deering
Mesquite, Texas

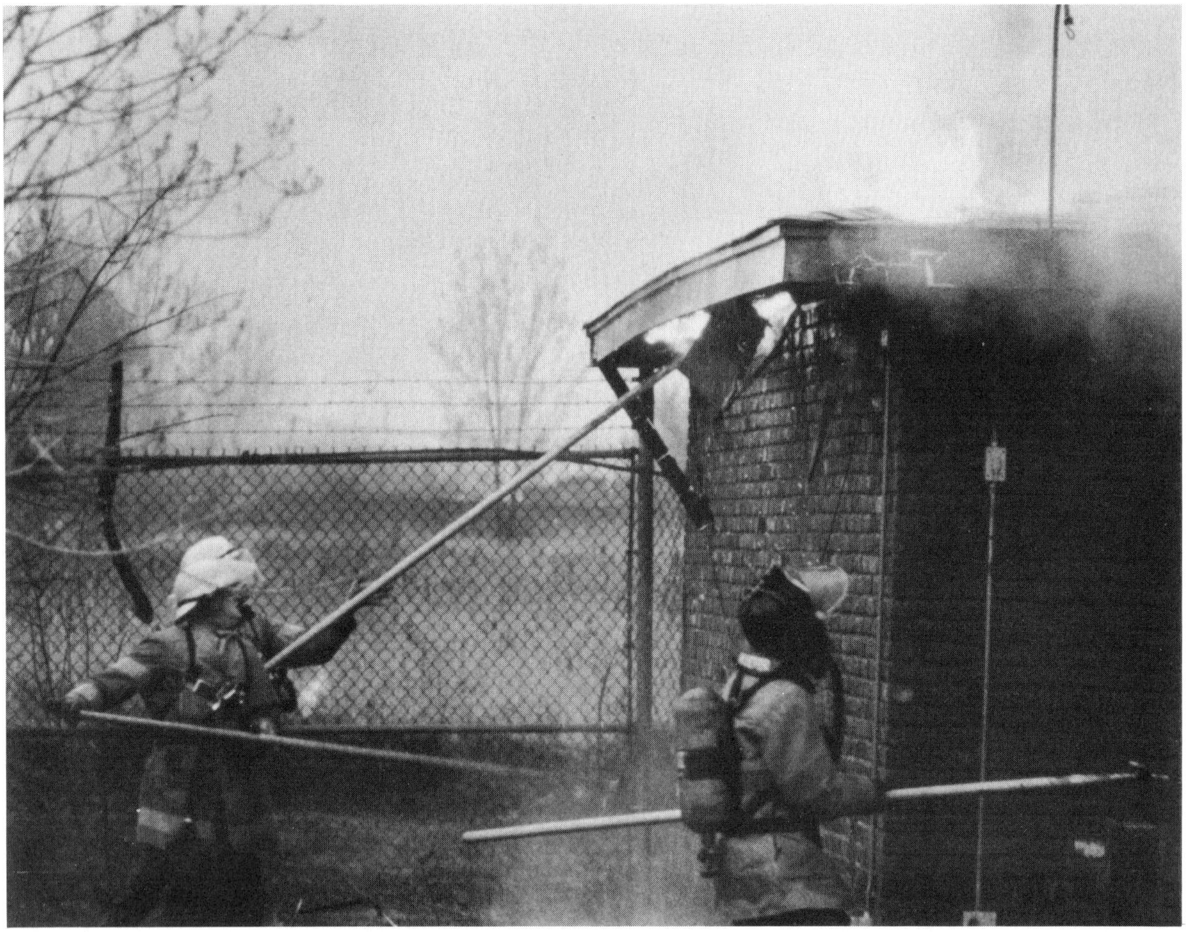

Opening Up
Peter A. Danzo, IFPA
Ridgefield Pk., New Jersey

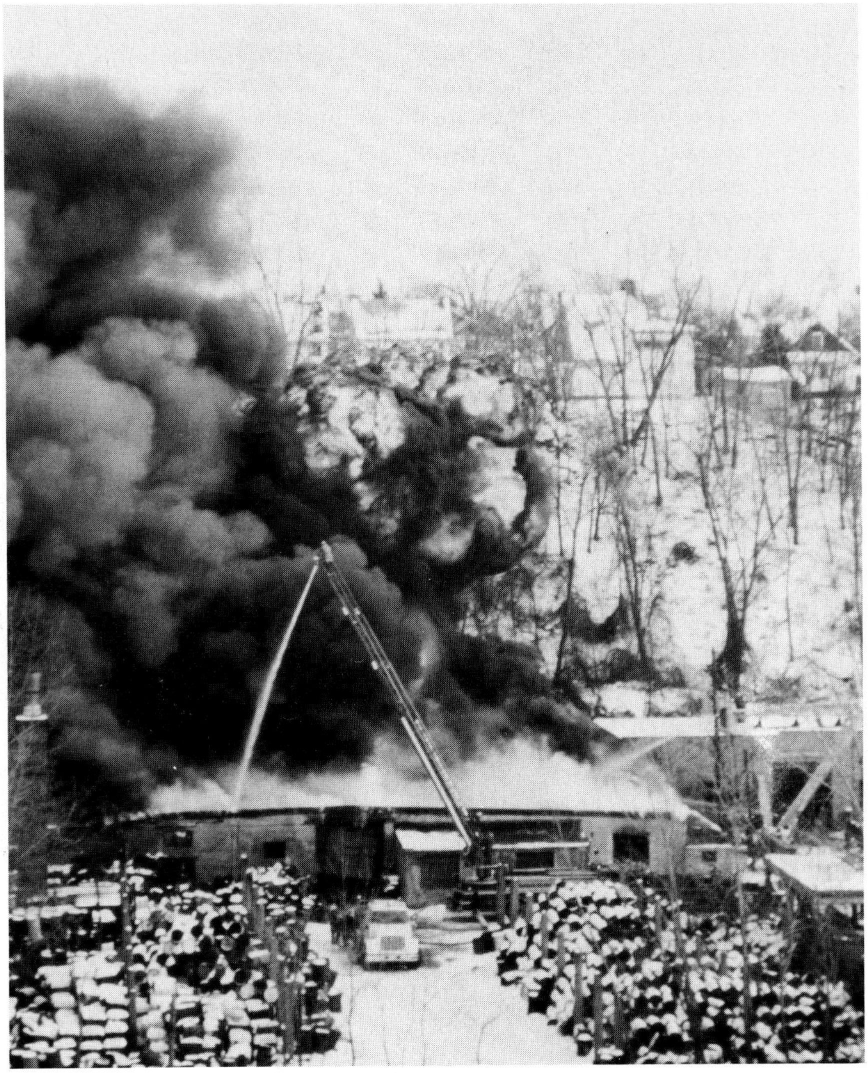

Barrel Plant Blast
John McCown
Lakewood, Ohio

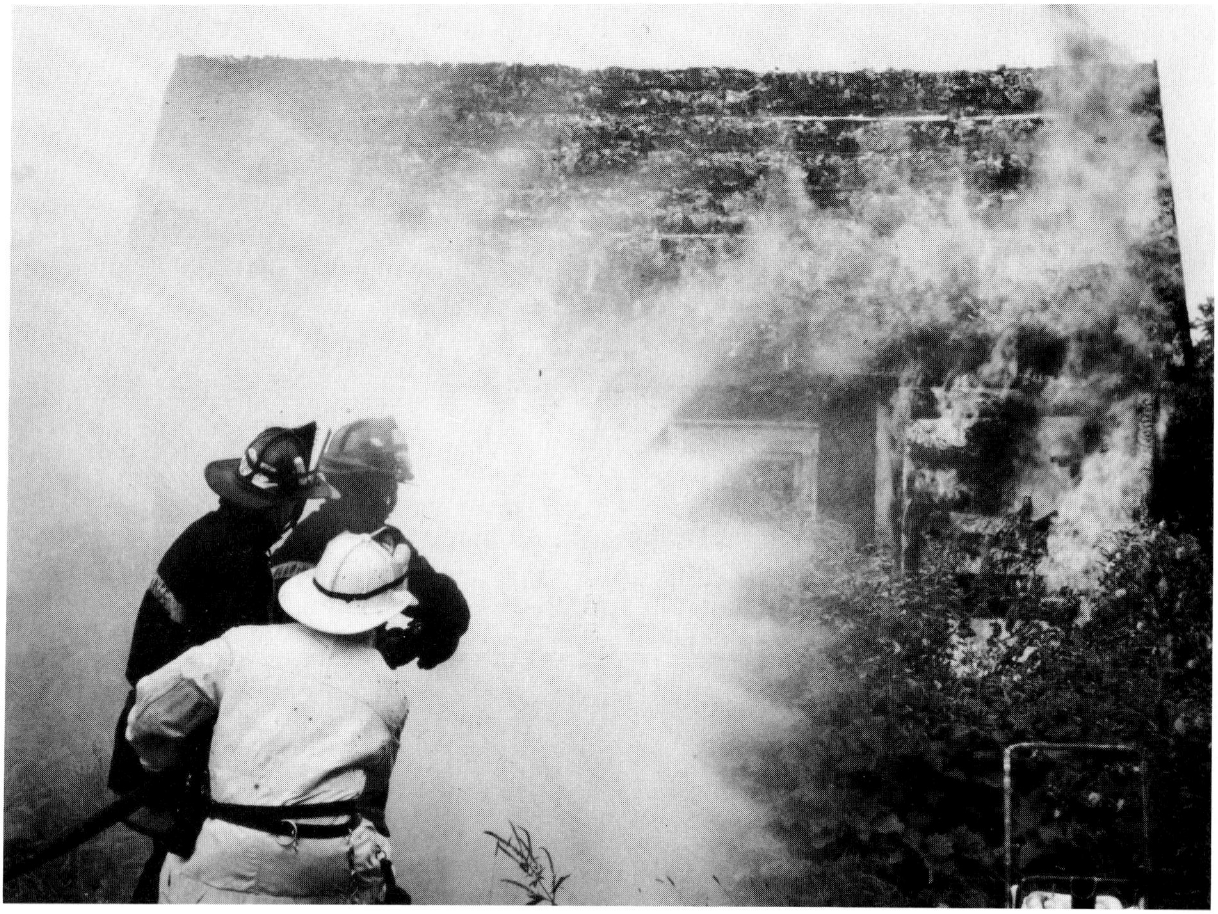

Fog Pattern
David J. Douglas
Racine, Wisconsin

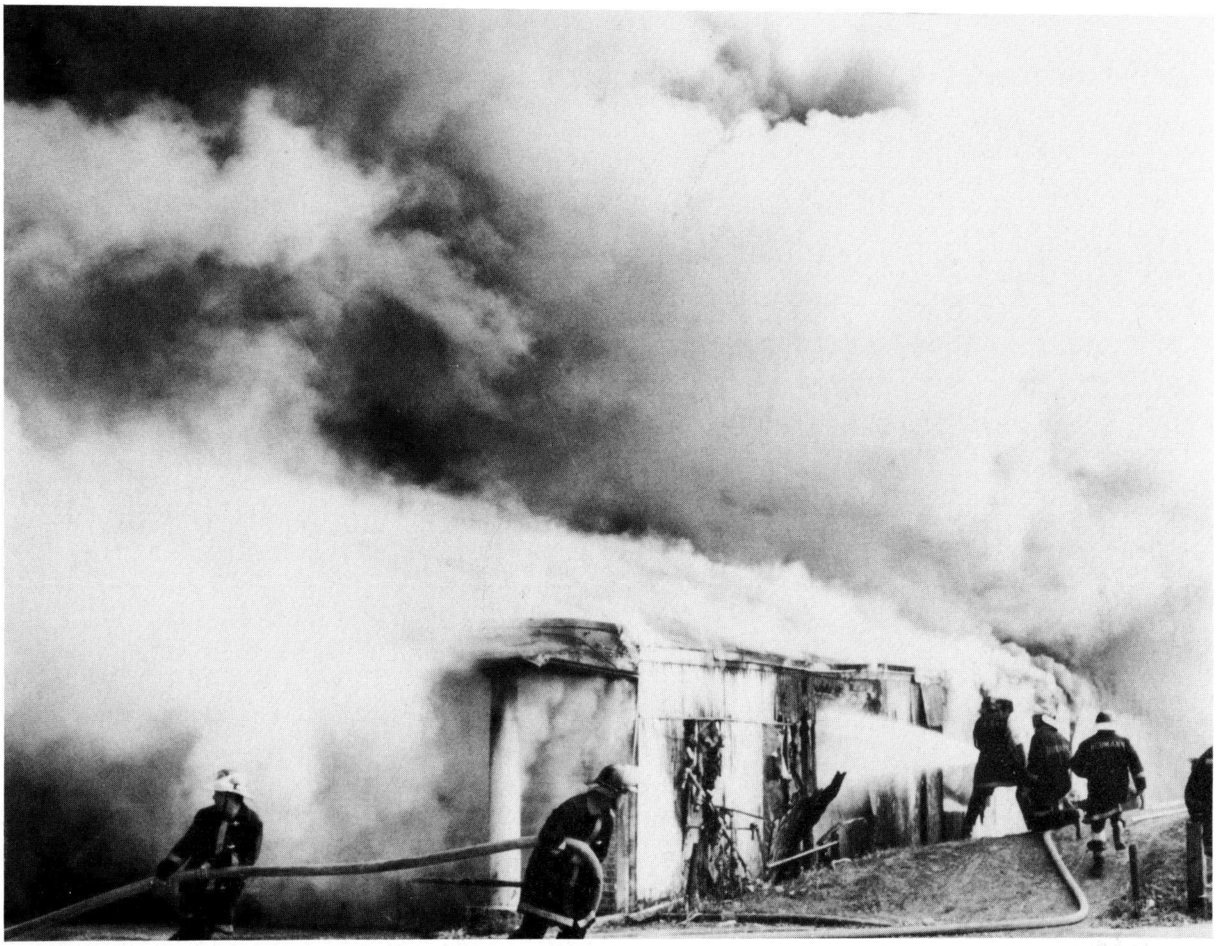

Chris Mickal
New Orleans, Louisiana

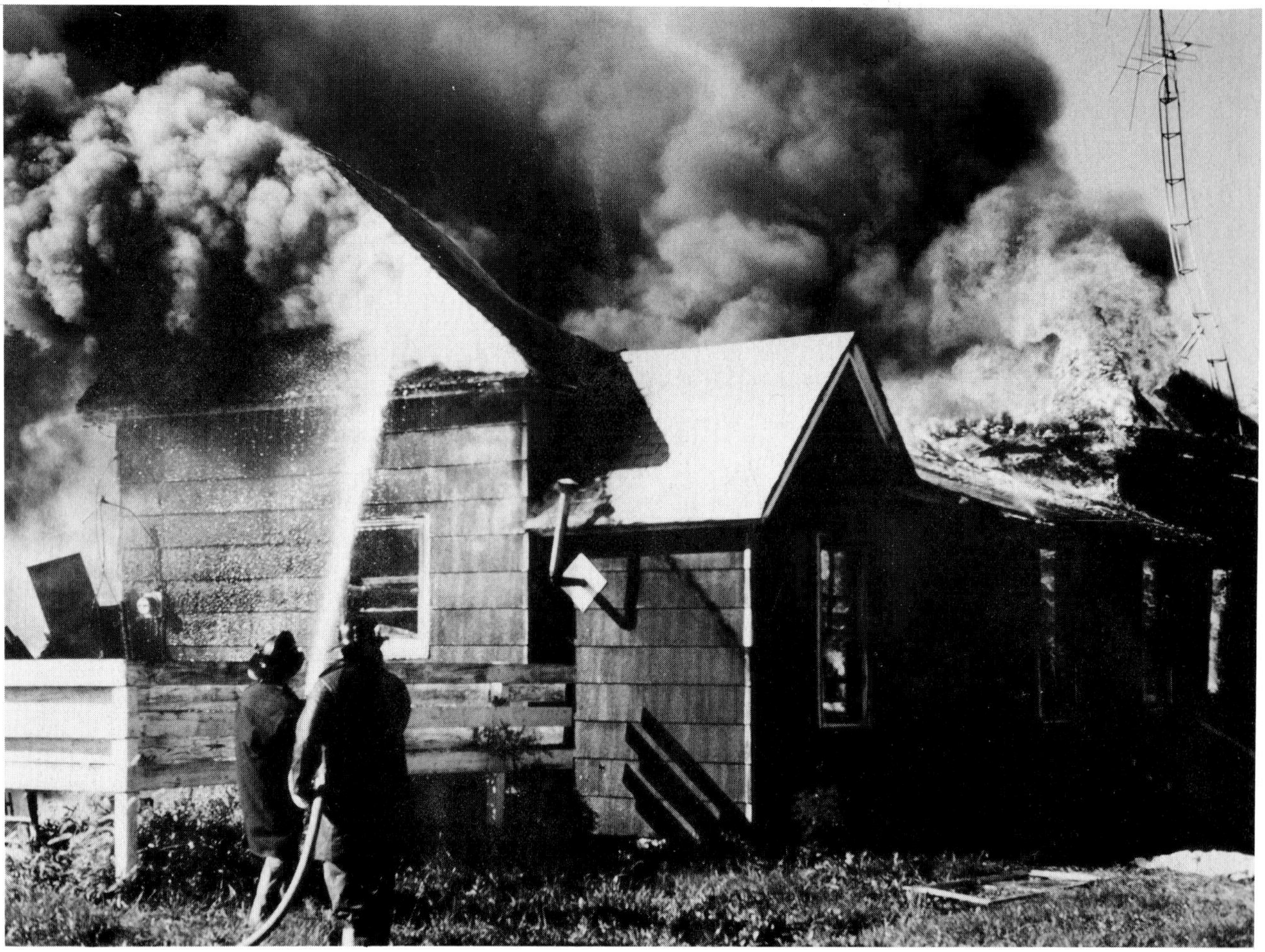

House Fire
Ronald S. Manwaring
Murphysboro, Illinois

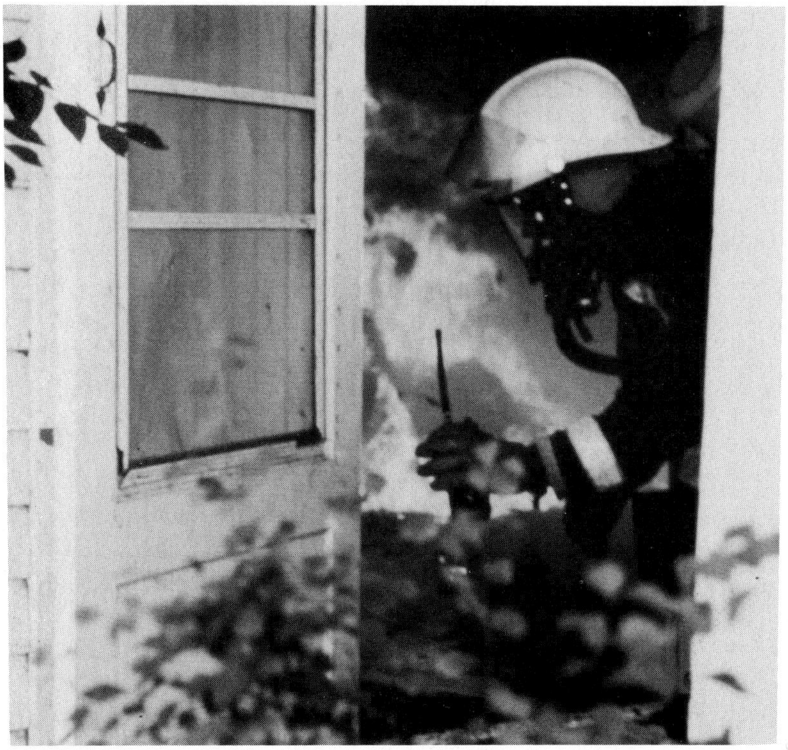

Red Hot
Kay E. Cermak
Normal, Illinois

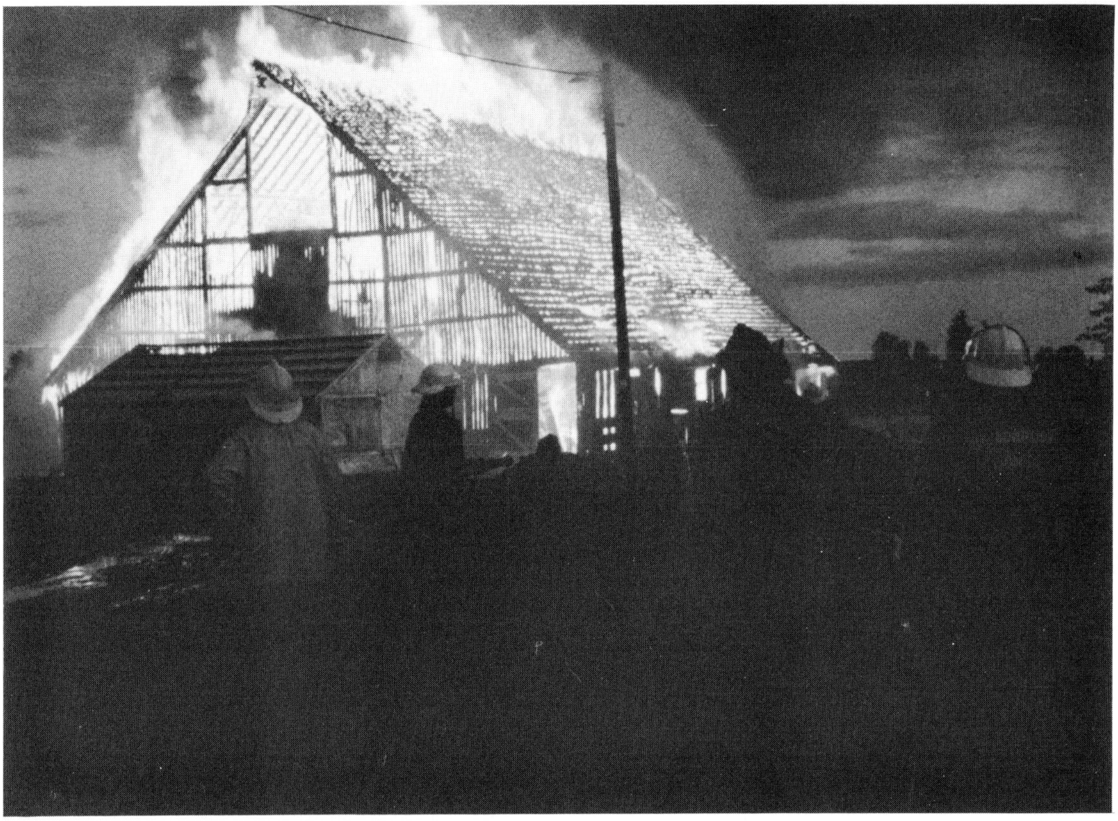

Untitled
Ernest J. Hansen
Gladbrook, Indiana

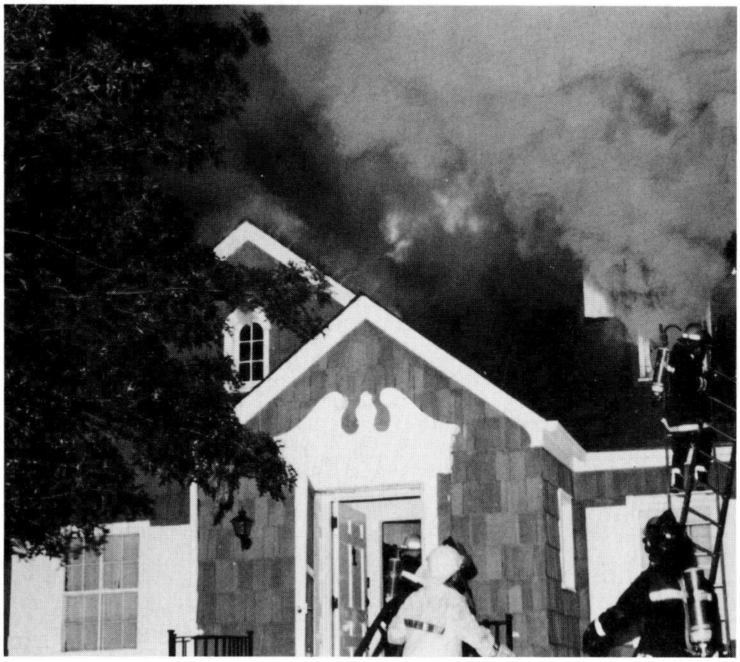

A Stubborn House Fire
Robert Wiesemann
Belleville, Illinois

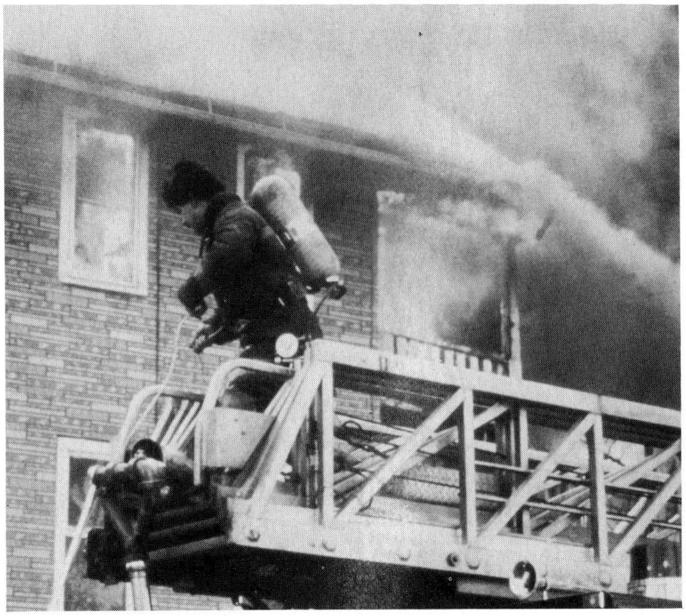

5 Frames On The South Side
James T. Regan
La Grange Park, Illinois

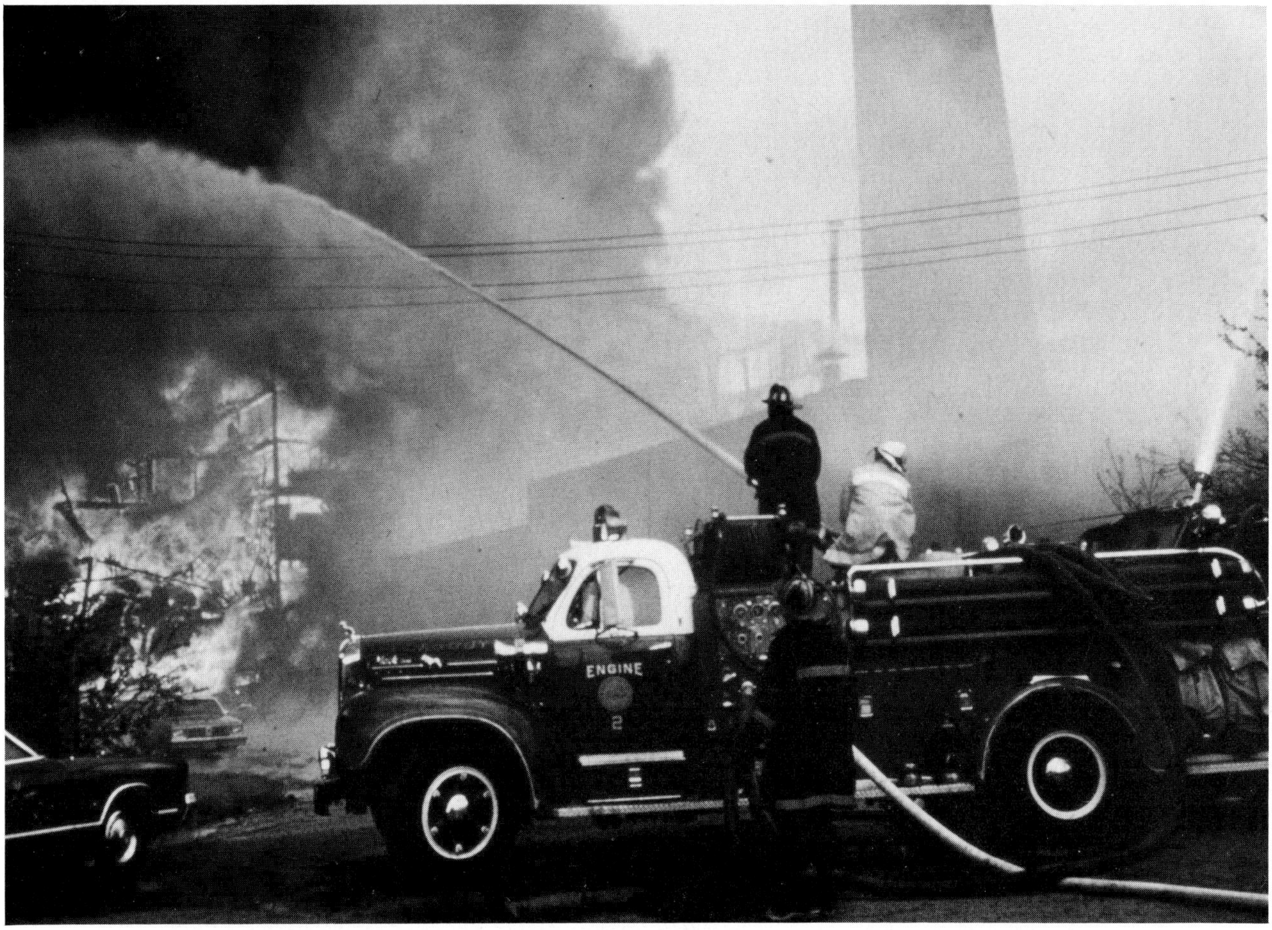

Sometimes There is Never
Enough Water
John Barry
Lynn, Massachusetts

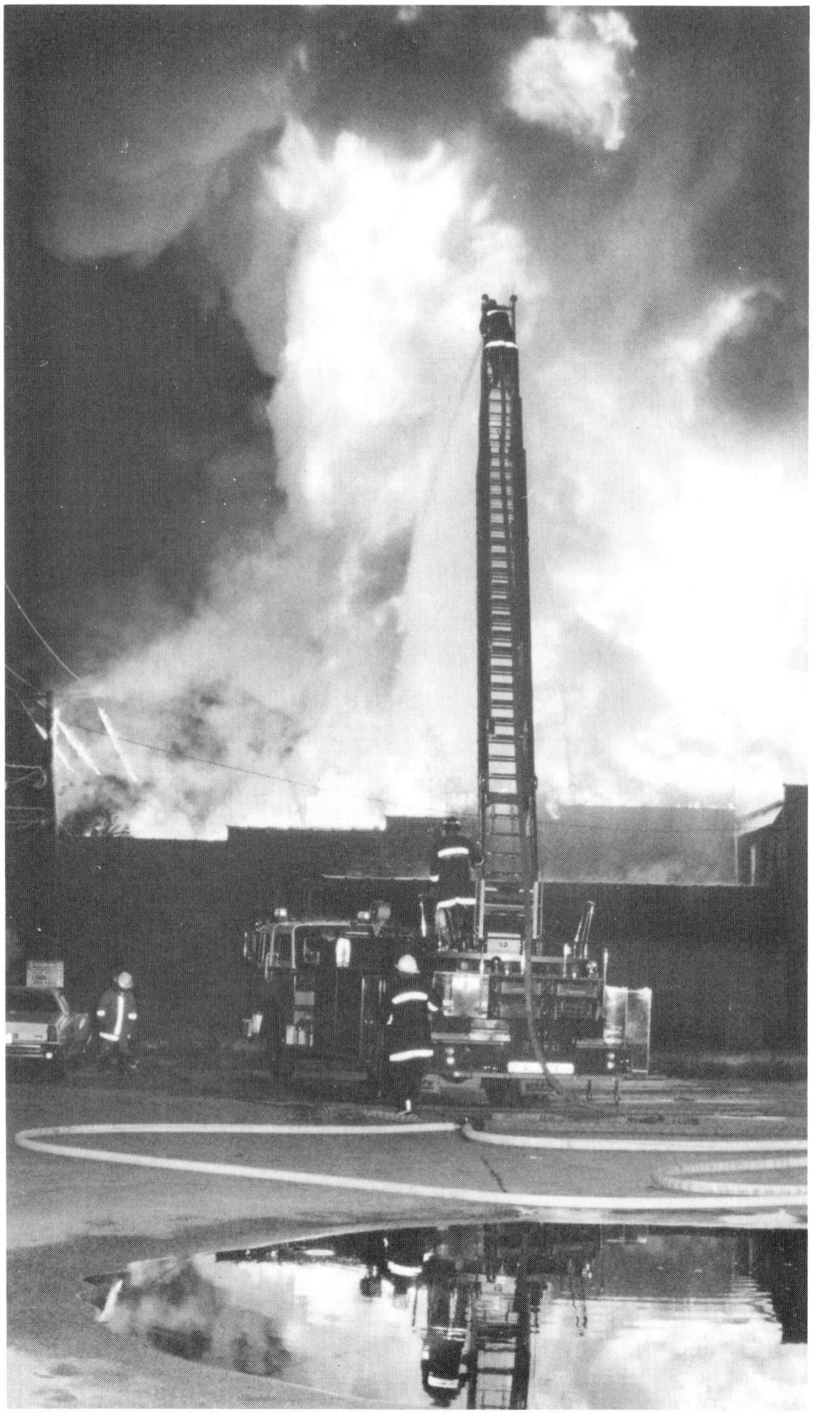

I Think It Just Went Through The Roof
Gordon J. Nord, Jr.
Lyons, Illinois

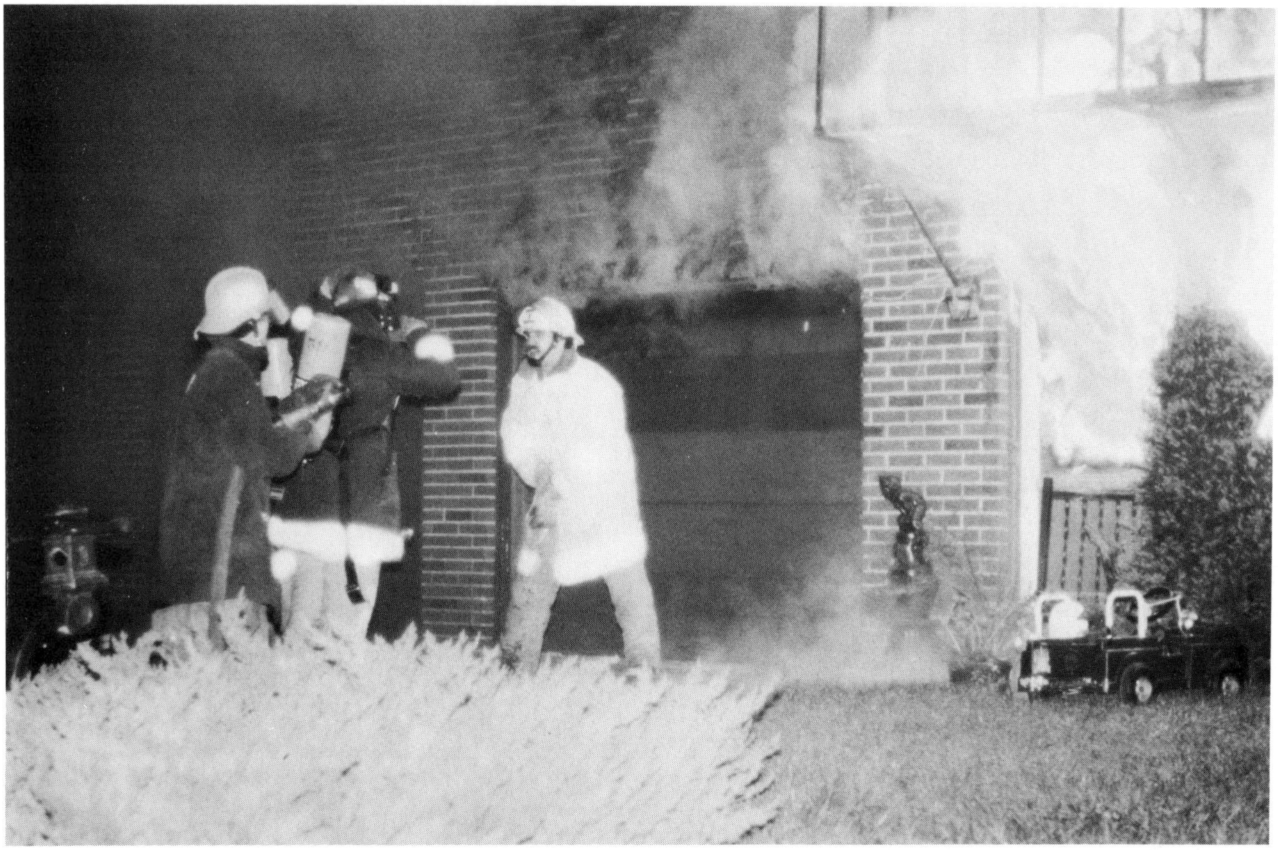

Patrick Grace
Binghamton, New York

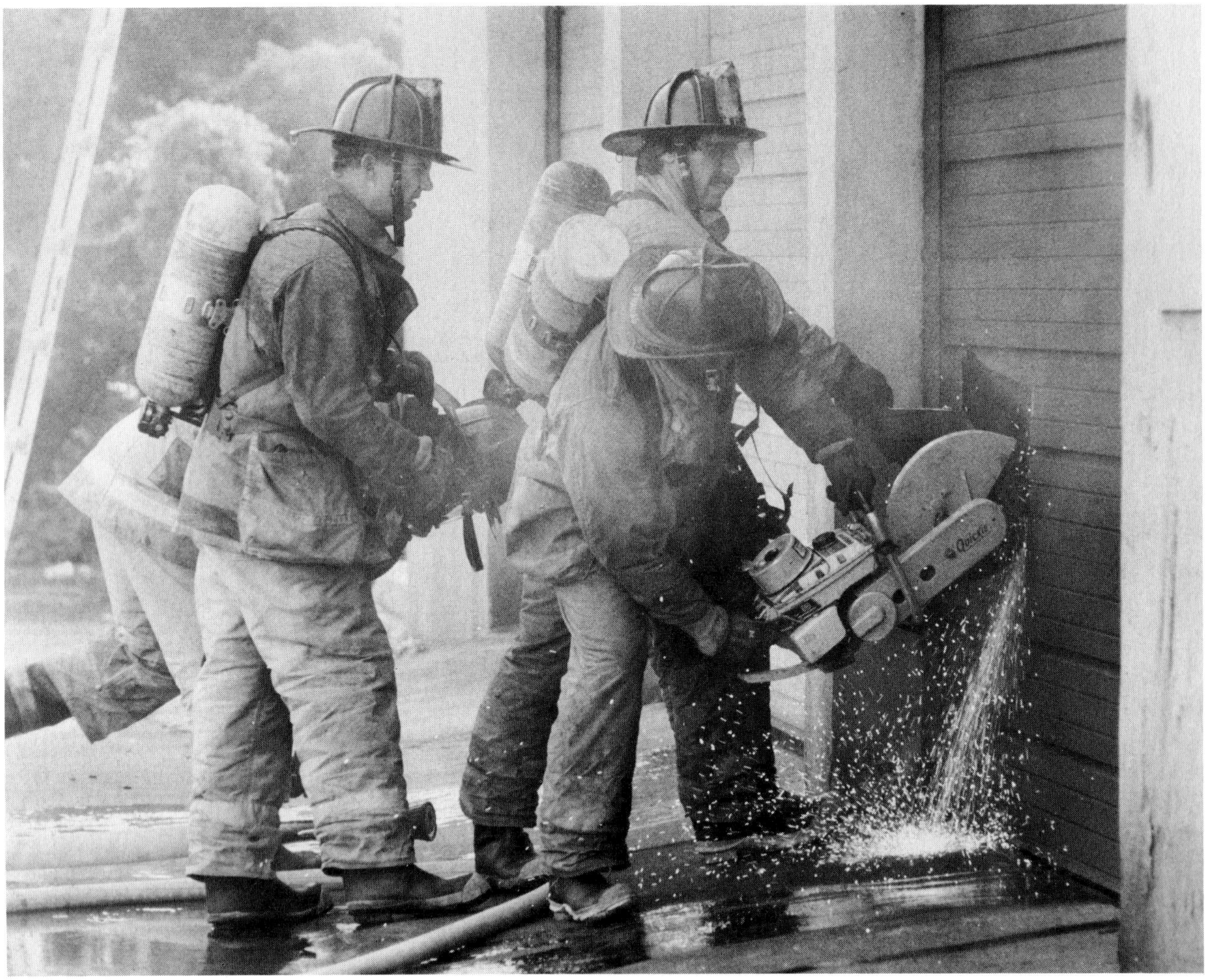

Truck Company #1 "Opening Up"
Fort Lauderdale News and Sun-Sentinel
Ft. Lauderdale, Florida

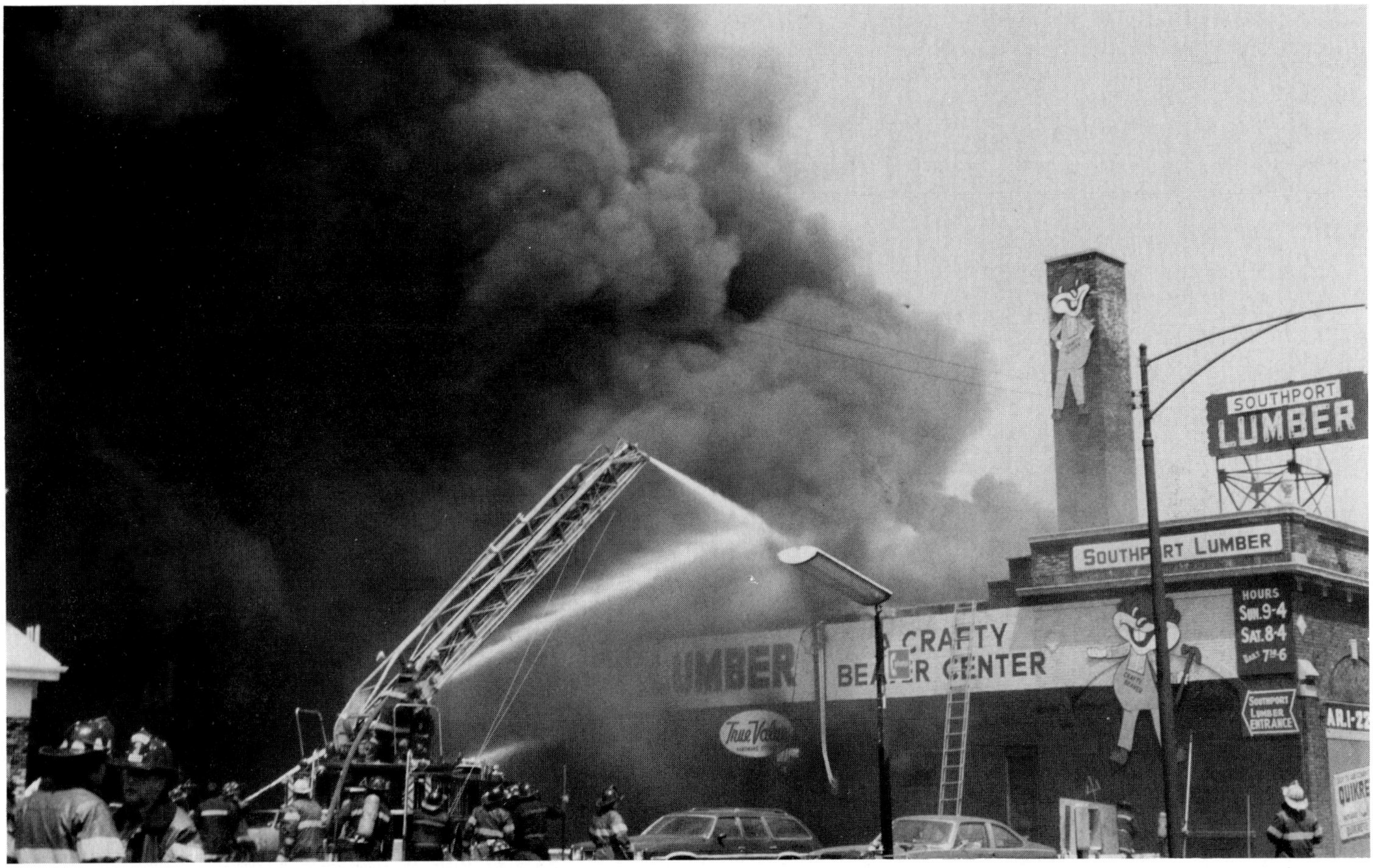

Going To Work
Charles Bleck
Chicago, Illinois

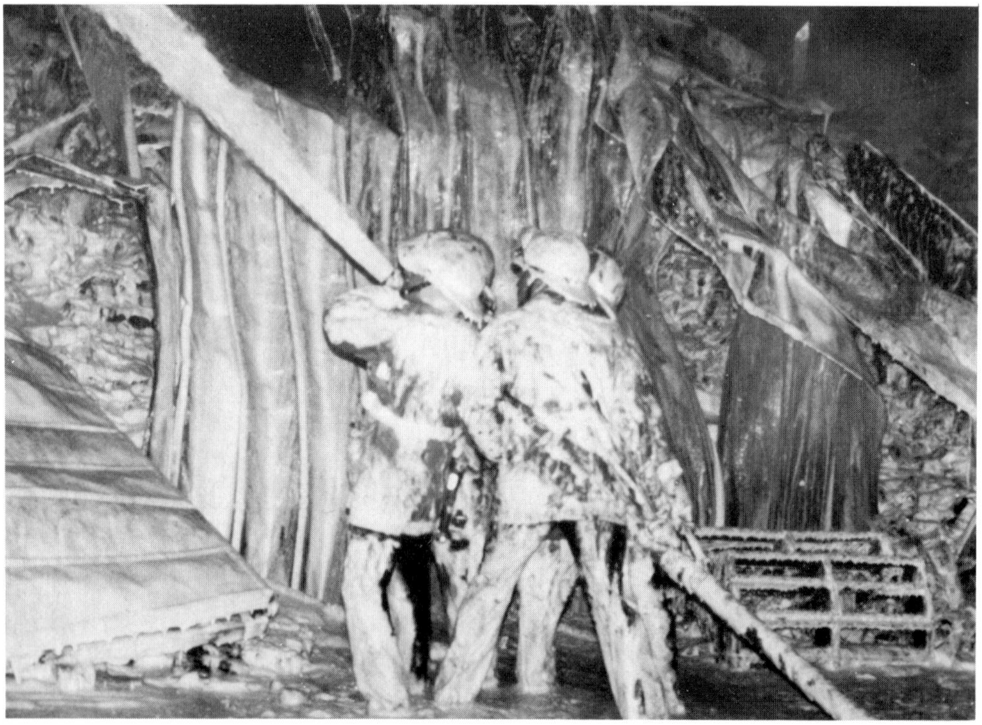

Frozen in Foam
Jill Meryl Levy
Cupertino, California

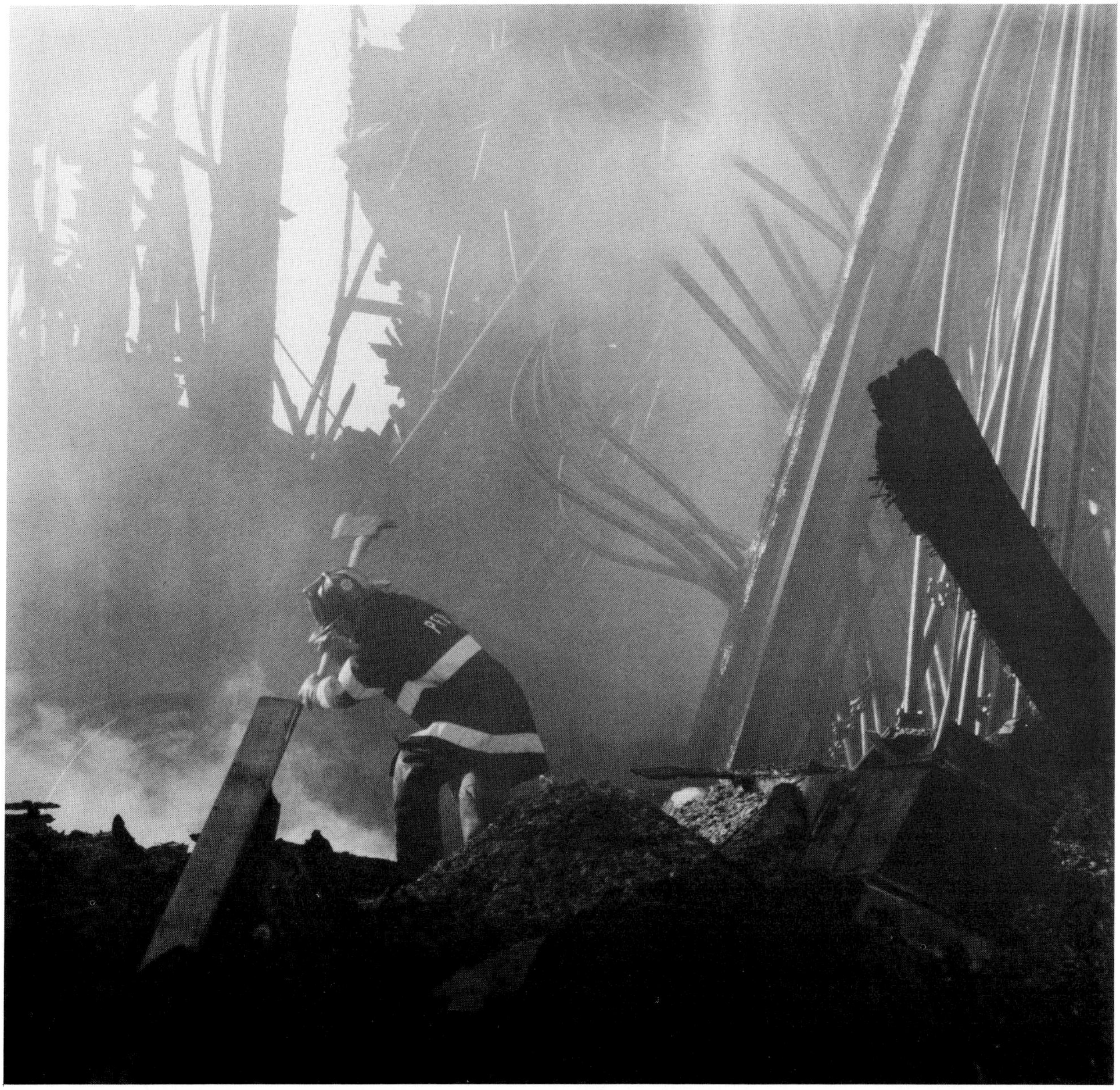

All in a Day's Work
Robert M. Proudhomme
Oceanside, New York

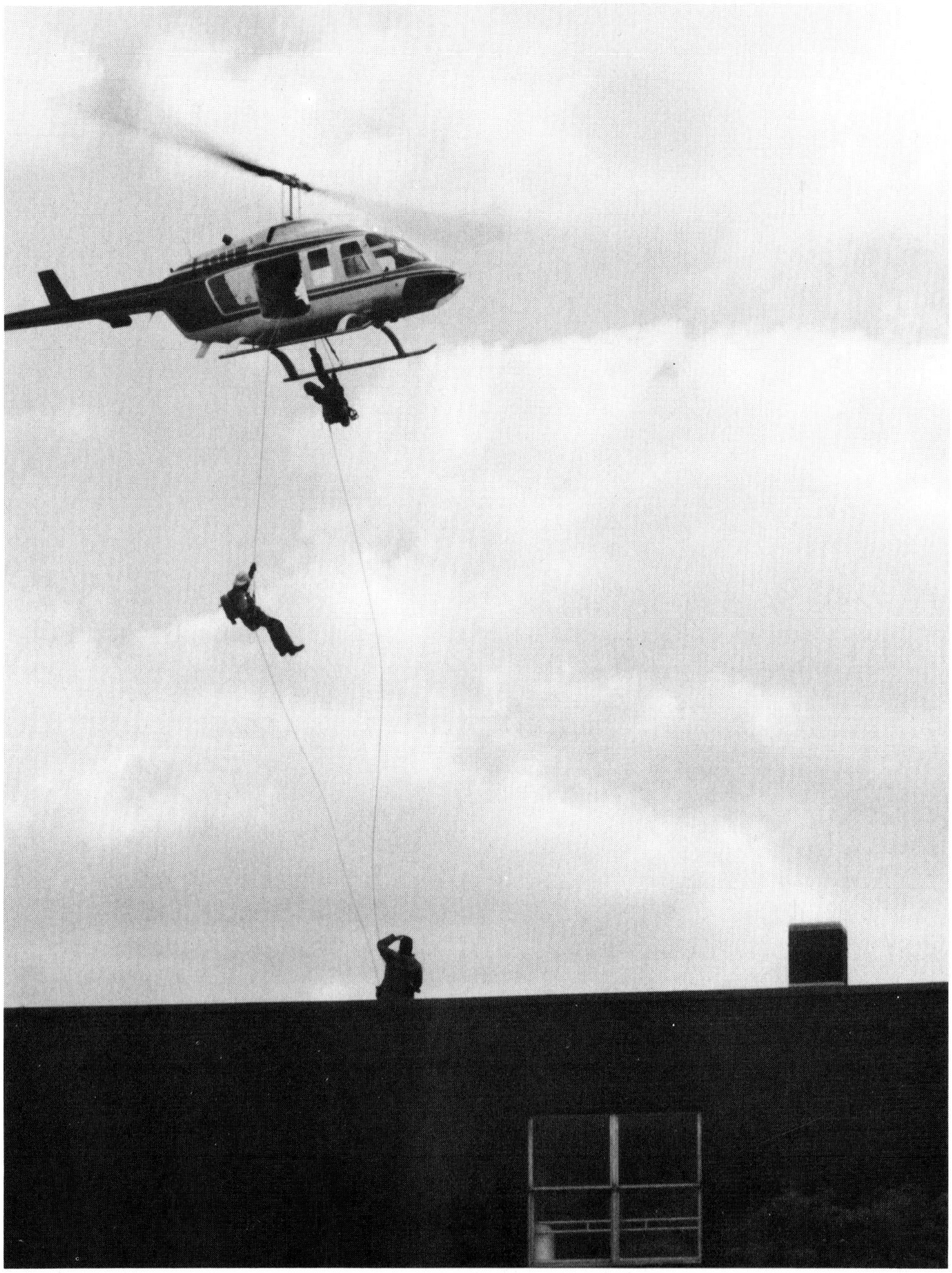

Things Can Easily Go Wrong
R.R. Thomson
Upper St. Clair, Pennsylvania